AF765957

September 1992, London

The figural and the literal

The figural and the literal

Problems of language in the history of science and philosophy, 1630–1800

Edited by Andrew E. Benjamin,
Geoffrey N. Cantor *and* John R. R. Christie

Manchester University Press

Copyright © Manchester University Press 1987

Whilst copyright in the volume as a whole is vested in Manchester University Press, copyright in the individual chapters belongs to their respective authors, and no chapter may be reproduced whole or in part without the express permission in writing of both author and publisher.

Published by Manchester University Press
Oxford Road, Manchester M13 9PL

British Library cataloguing in publication data

The figural and the literal: problems of language
in the history of science and philosophy, 1630–1800.
1. Philosophy, European – History 2. Linguistics
– History
I. Benjamin, Andrew E. II. Cantor, Geoffrey N.
III. Christie, John R. R.

190′.14 B791

ISBN 0 7190 1486 7 *hardback*

Library of Congress cataloging in publication data applied for

Photoset in Linotron Sabon by
Northern Phototypesetting Co., Bolton

Printed in Great Britain at the University Printing House, Oxford
by David Stanford, Printer to the University

Contents

Contributors

Andrew E. Benjamin, Philosophy Department, University of Warwick.

Geoff Bennington, School of European Studies, University of Sussex.

Geoffrey N. Cantor, Division of History and Philosophy of Science, Philosophy Department, University of Leeds.

John R. R. Christie, Division of History and Philosophy of Science, Philosophy Department, University of Leeds.

Stephen Copley, Department of English, University College, Cardiff.

Tony Davies, Faculty of Arts, University of Birmingham.

Jan V. Golinski, Churchill College, Cambridge.

Michael Moriarty, Caius College, Cambridge.

George MacDonald Ross, Philosophy Department, University of Leeds.

Prefatory note

With the exception of the essays by Stephen Copley and Tony Davies, all the papers in this volume were presented at a conference held at Leeds University, 10–12 January 1984.

1 *John R. R. Christie*

Introduction: rhetoric and writing in early modern philosophy and science

The field of history, and more particularly the specialised branches of intellectual history, have been rendered increasingly problematic in the last two decades. Althusserians and *Annalistes* bring reconstituted versions of the past and its subjects to the fore, while Foucault has shifted our focus away from the individual producer and towards the principles and rules which produce specific discursive formations. Structuralist semiologies and post-structuralist theories of the sign, often in alliance with one or more of the foregoing, invade the territory of cultural and anthropological historians with confident and totalising ambition. While much of this remains opposed by the stubborn empiricist streak in Anglo-American historiography, Marxist and non-Marxist alike, and while from a less parochial perspective the novelties adduced above signify only a belated recovery by Anglophone academic culture of developments earlier initiated and carried through in France, these interpretive reorientations are now building up the possibilities of reconceptualising the past and our understanding of it. History, and how we write it, are questions forcefully reposed.

If these reworkings, multifarious in nature and centrifugal in tendency, can be said to have a common root, then it is a preoccupation with language. Whether we have to deal with Levi-Strauss's structures of mythological understanding, Lacan's version of the Unconscious, Foucault's notion of discourse or the arcane vocabularies of specialised scientific disciplines proposed by Kuhn, the uses, influences, complexities and difficulties of language move into focus. The present undertaking tracks this movement into intellectual history. For while the reworking of 'history' has opened up manifold new avenues, the purlieus of intellectual history, its categories and assumptions, remain relatively, if not wholly, under-examined. To emphasise the subject of language for intellectual history, in particular for the history of scientific and philosophical thought in its formative early modern phase, is the work of this volume.

Innumerable forms of the question are available, but the most immediately workable and fruitful site of engagement is an oppositional pairing which is simultaneously general to the whole field of language, yet specifically crucial to scientific and philosphical discourse. This is the distinction, to give its most common usuage, between the literal and the figural, a distinction which in turn underpins further analogous polarities: grammar/trope, truth/poetry, science/rhetoric, discovery/justification, philosophy/literature. From fifth century BC Athens down to the latest times, this distinction has a long, complex history. If this volume engages that history *in medias res,* it is because the early modern period witnessed strong and profoundly-held reorientations of the distinction, very largely as a result of the conjoined and inseparable revolutions in intellectual history constituted by philosophical rationalism and scientific empiricism. The distinction was thus built into modern philosophical and scientific sensibility from the outset, and continues to reinforce, not just perceptions of philosophy and science from without, but essential practices internal to those activities and their public presentations. By re-reading a selection of the canonical names and works of the early modern intellectual revolution, the essays which follow initiate a questioning of the validity of the literal/figural dichotomies and their unproblematised role in the history of modern Western thought.

Re-reading the canon: this may have a tendentious ring. But *reading* is exactly the point at issue. Conventional forms of reading science and philosophy only remain unproblematic within the tacitly accepted validity of the distinction. A less supine attitude attempts to realise reading as a much more challenging and interesting process, which recognises as problems, as areas of opacity, what otherwise remain transparent and invisible to interpretive understanding. Consider another pairing, related to the figural/literal distinction: that between content and expression. Conventional scientific and philosophical reading tends strongly towards discrimination by content. However, to focus on content at the expense of expression introduces another and comparable form of blindness, for the work of the expressive mode, language, will tend to go unglimpsed unless matched by an active reading constantly aware of the textual tactic and strategy which written language always and inevitably embodies. The suggestion is, then, that the primary encounter with any text, be it metaphysics, poetry or biology, is linguistic, for texts are made of language.

Thus baldly put, the suggestion sounds banal. Its heuristic point depends upon the willingness to follow through its implications. Following the historical defeat of rhetoric by philosophy, itself a part of the early modern intellectual revolution, philosophy and science pretended to a profound literalness of expression. The field of figuration, of rhetoric, of non-literal persuasion, is firmly fenced off, and large warning notices affixed. The assumption that conviction may be induced in the reader has as its premise the shared promise that non-literal expression, the resort of rhetoric, will not be practised, because it has been ruled out of court. The whole question of style and expression can therefore be dispensed with, to the extent that the philosopher or scientist under scrutiny has respected the imperative of literal clarity; to the extent that this imperative has not been respected, the text is read in spite of, or against, its style, such is the expectation created by the legislative reflex to marginalise notice of expressive modes when in quest of philosophical truth. This expectation can also be characterised as being about reciprocity between the nature, or subject, of the text and its style. The scientific or philosophical text is expected to employ language facilitating the unmediated expression of literal truth, as opposed to the literary text which, obliged to cajole the imagination into pleasure, will make great play with figural language in a field of indeterminate reference. These opposing expectations then also function as valorisations, for the descent of science to figuration is seen as 'excessive' and 'fanciful', while the more authentically mundane referentiality is contained in a work of literature, the less 'imaginative' it may be held to be.

This is admittedly an over-defined and caricatured rendition of the opposed legislative scenes and expectations of two types of reading. For a start, much traditional literary criticism has worked with a form of literary reference as unproblematised as philosophy's conception of literal reference, and is thereby open to the kind of critique envisaged here for science and philosophy. Nevertheless, caricatures can and do serve a proper purpose, revealing by over-emphasis essentials of the subject. It is indubitable that, internal to the practices of science and philosophy themselves, such words as rhetoric, fancy, metaphor, at times 'imagination' itself, are critical, at times abusive, terms, and, as early as the second half of the seventeenth century, had come to form a recognisable rhetorical system of stigmatisation, particularly at the hands of the English empiricist scientists. Another possible and plausible reaction would show that philosophical and scientific writing cannot be caricatured in this way. It could show how such writing uses analogy and metaphor, often adroitly, as aids to clarity. From there, it might claim that rhetorical usage, whether implicit or explicit, has its place in such discourse, given proper control and understanding, and that scientists, philosophers and their historians recognise this. Yet such a response immediately concedes the crossing of a consistently and heavily policed border, and it is such transgression, undertaken in pursuit of the very goal the border exists to protect, which forms one possible starting point for the kind of enquiry to which our enterprise is devoted. Rather than live uncomfortably within the logic of a paradox which simultaneously insists on the literal/figural distinction, inscribes the essential nature of science and philosophy around the pole of the literal, yet acknowledges the figural in pursuit of the literal, it is sensible instead to entertain the notion that the literal and figural are part of a common linguistic field, and demand a simultaneous investigation. Once this notion is installed as a conscious strategy of reading, then the literal-figural opposition ceases, in principle at least, its overt and covert structuring of intellectual historiography.

Radical gains may also ensue, for it is clear that if science and philosophy can be read consistently as rhetoric, and not merely as containing marginal and negligible rhetorical elements, then our understanding of those activities, and then through them the worlds they claim to reveal and represent, should be profoundly altered. Some recently completed studies foreshadow the form of this alteration. Owen Hannaway's study of *The Chemists and the Word* (Baltimore and London, 1975), undermines the usual historical assumption of the constancy of scientific language by paying scrupulous attention to the linguistic surface and textual strategy of the works he examines. In so doing he is able to further our historical understanding of how and why poetic and literalist modes of representing the natural world began to separate early in the seventeenth century. He does this without conceding any priority or privilege to the literalist mode, but instead indicates that it is a historical formation as contingent, and as rhetorically implicated, as the poetic chemistry it sought to supplant. It should prove possible to undertake studies which will perform comparable work upon the whole literary text of 'experiment' introduced by seventeenth-century scientific empiricism.

If we move instead to examine contemporary work in the philosophy of science, Kuhn's description of how scientific texts and text-books function in the formation and furtherance of paradigmatic forms of scientific practice offers as likely a starting point as any. Kuhn's description is often read politically: a scientific text-book is an authoritative, perhaps authoritarian device which instills acquiescence to paradigmatic commitments in the reader. To notice this political dimension should also be to raise the question of how, exactly, might a text convey such authority; and only a rhetorically and discursively sensitive reading will be able to provide the answer.

If these have the air of randomly chosen examples, it is because the installation of rhetorical reading does not entail a single method or tactic of working, nor, historically speaking, a particular starting point. That said, the essays presented in this volume do offer certain narrative and thematic continuities, and illustrate approaches which have convergent as well as divergent tendencies.

The essays on seventeenth-century figures re-examine basic elements in what historiography has always seen as one story, usually called 'The Scientific Revolution'. 'The Intellectual Revolution of the Seventeenth Century', or some other such name. What the authors emphasise is the linguistic dimension of that story, and by concentrating on language as both the conditioning form and the (often-ignored) subject of the story, the narrative of early modern intellectual change, which is too often the historiographical validation of the change it describes, begins to be challenged and altered. Andrew Benjamin's essay on Descartes does not question the methodological foundation of Cartesian epistemology, but instead reveals how, as a text, the *Discours de la Méthode* questions itself. Descartes' efforts to promote his methodology in autobiographical terms involve a contradictory complex centred on the notions of resemblance and representation. On this reading, the *Discours* becomes a drama of non-reconciliation between the fictive self encoded as narrative fable and the knowing subject encoded as logic, each bound to perform at variance with the other, but with logic always requiring the story of its emergence to make its way in the world. Benjamin's Descartes tends to appear as much the progenitor of the novel as of modern philosophy and science. George MacDonald Ross's essay on Hobbes, by contrast, resolves a potential discord in Hobbes's view on language and meaning. The two different theories obtained from *De corpore* and *Leviathan* turn out to depend, not upon alternative conceptual priorities, but upon different socio-textual locations. The technical linguistic signs of philosophy require extensive notation (active mental registration) to fix meaning, whereas the signs of common language have meaning pre-notated by the usages of history, custom, etc. This division, locatable also as between private and public in Hobbes's texts, therefore specifies two modes of meaning whose social nature and performance differ significantly, but whose theoretical statements do not require an oppositional ordering of note and sign.

These themes, of the status of the self's testimonies and narratives, and of the social and conceptual order of signification, reappear in Jan Golinski's and Tony Davies's essays on Robert Boyle and John Wilkins. Golinski sees Boyle's *Sceptical Chymist* as an instance of oppositional rhetoric, where the sceptical dialogue emerges as the extended narrative figure through which a new science of matter sought to invalidate more traditional discursive orders. The positive discursive strategy which accompanied this critique, the experimental essay, relied on experimental narration supported by the authenticity of the narrator and by creditable witnesses. Boyle's close colleague Wilkins shared with him, and with Thomas Sprat, in the early Royal Society of London's attack on rhetoric in general, but it was Wilkins, of all the *virtuosi,* who attempted to push a new linguistic ordering of the world to the limit. As Davies shows, Wilkins' scheme for a 'Real Character' was total in its scope and utopian in its epistemological desire for unity of sign and referent, the ultimate Hobbesian notation. Locke, usually and reasonably seen as some kind of culmination of seventeenth-century philosophy's drive against the rhetorical, is submitted by Geoff Bennington to an inquiry which tracks his obvious polarities, between philosophy and rhetoric, truth and the world, back through his views on language to the core of his epistemology; here is seen now not the *tabula rasa,* but an already differentiated entity, so that the Lockean subject, cured as it might appear of rhetorical abuses, nonetheless contains within himself the original form of the condition from which philosophy set out to deliver him.

The interpretive and historiographical significance of each of these essays expands considerably beyond the indications offered by such brief introductions. The collective suggestion of Benjamin's and Bennington's essays is that early modern philosophy's attempt to create *homo sciens* could not finally survive within the linguistic divisions constitutive of the attempt. This suggestion philosophically frames, we might say, the Royal Society's motto *Nullius in Verba,* which, as Golinski's reading of Boyle's rhetorically discursive strategies and Davies's appreciation of Wilkin's utterly scripted universe strongly imply, should now be inverted – *Totius in Verba.* If this formulation proves usable for further application, and there is no reason to believe the contrary, then many of the conclusions and methods of conventional intellectual historiography, already in defensive posture against sociological attack, will prove even more untenable.

As the volume moves into the eighteenth century, its focus becomes more selective within the expanding range of philosophical and scientific activity. Geoffrey Cantor's essay on the evolution of key terms in optical science shows the ways in which the nature of those terms mutates as they move through time and a various discursive space. The fixing or unfixing of terms as parts of propositional or metaphorical discourse is here seen as a historical process, a significant methodological point if such studies are to be undertaken successfully in a fully historical form. Michael Moriarty's study of Diderot examines another basic axis of metaphoric creation through Diderot's sensuous epistemology, the comparative and substitutive transferences of terms from one sensory set to another (e.g. from touch to sight). Diderot's synaesthesia, Moriarty shows, was the necessary metaphorical basis for his general philosophical materialism. The project of materialism is a sequence of metaphors, with machines playing an increasingly featured role in the process. The enterprise of philosophy and science is cast back once more upon the resources of figural language, in this case as apparently the only means of giving materialism an intelligible grounding against idealism.

The culture of mid-century Enlightenment receives further attention from Stephen Copley and John Christie. Copley's reading of Humean ideology shows how ideology itself is not simply to be read off particular concepts, or even off concepts which contradict.

Hume's theorisation of culture both invites the reader's complicity by the assumptions inherent in its literary form, the polite essay, and reproduces consistently unsignalled changes of meaning whose net effect, if not carefully read, is to conceal basic social distinctions, which his colleague and compatriot Adam Ferguson then consistently brought to light. Christie's essay attempts to recover Adam Smith for intellectual history by demonstrating his interests in science, epistemology, language and rhetoric, areas whose intellectual history generally take little notice of Smith. The twin themes of materialism and the figure of the machine which preoccupy Moriarty's essay also receive notice here, for Christie suggests that Smith's theory of language is just one application of a general analytical model, at once historical and mechanical, present also in Smith's treatment of science and economics. At the same time, Christie uses interpretive frameworks derived from Derrida and de Man to reveal the ways in which this second Adam, whose original sin is usually seen as political economy, is first implicated by the snake, language, within the stressed antinomies of nature and history, grammar and rhetoric.

This introduction has emphasised narrative, thematic and methodological continuities present in the contributors' essays. Concludingly, an opposite and minatory emphasis upon the implications underlying certain equally present methodological divergences is also warrantable. Perhaps the main distinction is to do with the degrees and kinds of historicity evident in this collection. Some of the essays are more firmly embedded in historical reference than others. Texts, in other words, appear in two forms: within a historical context variously constituted by a number of extra-textual features; and within a present texture, or inter-text, whose centre-piece is the examined text but which can include elements from the farthest past or near present if their interpretive or analytical presence is necessary or relevant. This, it should be immediately pointed out, is no definitive categorisation of the volume's essays. It points to a tendency, but the methodologically interesting point is the way in which some essays at least mingle their textual presentations. Davies's treatment of Wilkins, for example, uses Derridan concepts to understand Wilkins, but is also firmly committed to a historically specified explanation, or potential explanation, for Wilkins' Derridan writing. Christie's essay also combines a historical concern to sketch an intellectual and ideological context for the emergence of Smith's views with a reading of Smith derived from theories, part of whose power is that they do not oblige themselves to care whether Smith wrote in the eighth or the eighteenth century, for they are general theories of texts, of writing itself, seen as a para-historical condition. It would genuinely seem then, that for certain purposes there is a reasonable alliance between these contrasting ontologies of the text. Deconstructive insights can be given plausible historical locations. If deconstruction adds some historical sociology, if historiography is more aware of textuality, they may join hands across the gulf of reference. This is very probably a necessary movement in the present state of interpretive enquiry, but its theoretical and methodological legitimacy may ultimately have to be questioned. Historiography cannot do without a referentially reliable past tense, and denominative order, whereas deconstruction has no need of, and must certainly hesitate to provide a justification for, any such things. The latent methodological point of this book may then finally be that it works historically to displace the kind of polarity instanced at the outset of this introduction, the literal and the figural in philosophy and science, only to find it replicated within the general concept of historiography itself. That though is a problem only adumbrated by this book, and may be postponed for the sake of reading it.

2 *Andrew E. Benjamin*

Descartes' fable: the *Discours de la Méthode*

2.1 *Introduction*

Philosophy conceives of itself as continually involved in a search for truth in which truth is envisaged as an end in itself. This self-conception is, of course, usually mirrored in the expectations held for philosophy. The search, evident throughout the texts comprising philosophy's history, is for our concerns more than adequately exemplified in the title of Descartes' philosophical dialogue. *La Recherche de la Vérité*. However despite philosophy's proclaimed goal being the uninterrupted enactment of a will to truth, it is our contention that this is in fact the articulation of a different and indeed more subtle strategy, one which could be described as a will to overcome error, or perhaps to be more precise, as the will to overcome the possibility of deception. Part of the aim of this paper is to trace both the unfolding as well as some of the consequences of this strategy.[1]

If truth is no more than that which ensues from a strategy that establishes two inherently incompatible domains – one being the domain of error and characterised by temporality and the other the domain of truth that is turn characterised by universality – then truth itself has the appearance of remaining unquestioned and unexamined in regard to the truth of truth. If truth is the result of a will to overcome deception then this leaves unposed, albeit almost as a matter of necessity, the question of the true *qua* true.

Philosophy's search for truth can be described, therefore, as involving these two interrelated components, i.e. truth as the consequence (but no more than the consequence) of having overcome the possibility of deception, and the unexamined status of truth. This is of course not to deny that the various ways in which truth is thought to be established are not themselves always the subject of constant scrutiny, rather it is to suggest that what remains unexamined is the essence of truth. It is as though the status of the true *qua* true was simply a given.

There are three areas for further investigation that are delimited by this particular construal of philosophy's concern with truth, and even though they vary in their relevance for a study of Descartes' *Discours de la Méthode* it is nonetheless advisable to outline them.

Accepting the assumption that the ascription of truth signifies the claim to have overcome the possibility of deception, and in so doing to have driven a wedge between the domains of error and truth, then the first area of investigation must concern whether or not the possibility of deception can be continually overcome. Furthermore, can these two domains be rendered incommensurable, as they would have to be if truth is to be both secure and secured? The possibility of an intermingling, let along an interdependence between them, would mean that the sign of truth was no more than the site of truth's self-deception. In other words truth could be based on that which could bring about the destruction of the basis of truth.

The second area of inquiry stems from the fact that, as has been mentioned, the status of the true *qua* true (i.e. truth itself)) is left unresolved. The problem and thus the area to be explored concerns how to interpret this unposed question. Probably the most intriguing response comes from the work of Heidegger, in which it is consistently argued that the refusal of Western philosophy to raise and thus to confront the question of the essential marks a continual and significant absence which does itself reveal the way in which philosophy has worked hitherto. This position is argued in 'Vom Wesen der Wahrheit'[2] in relation to truth; however the absence of the essential as an area of inquiry is discussed in texts as seemingly diverse as 'Vom Wesen des Grundes'[3] and *Nietzsche*.[4]

In the suggestion that this absence must be examined Heidegger is committed to the position that, not only do concepts have an essence, but furthermore, that the unfolding of history has obscured this essence, and hence he views their recovery as one of the tasks incumbent on philosophy. Whatever viability Heidegger's position has, it does of course lie in there actually having been an original, and now thwarted, will to truth. Part of the aim of this study of Descartes is to call the basis of Heidegger's position into question, for another response to Heidegger's claim that the essential remains unquestioned would be to argue that this is no more than the inevitable consequence of the way in which truth is established. While the force of Heidegger's observation is not in doubt, it should not, as does in his case, give rise to a search for lost essences driven on by a redemptive nostalgia, but rather it should be recognised that the question of the essential cannot be posed since all that divides truth and falsity is their incommensurability. There is nothing called truth (or falsity) that has a discreet identity and therefore of which there can be an essence. The search for essences is just one more example of philosophy's sedulous fascination with origins.

The third domain concerns what may be loosely called the ethical. An ethics based on the possibility of avoiding error can only be expressed in terms of imperatives. If the domain of error (and thus sin) is the world of lived experience, then it is quite clear that the ethical can be neither based on nor grounded in the world. The guarantee of the ethical will have to be transcendental in nature as it is only the transcendental which, in being beyond the domain of error, is itself beyond reproach. This sketch indicates that the ethical is embroiled in the same problem as truth. In other words, the problem alluded to earlier, namely the question of whether or not the epistemological division between the temporal and the transcendental can be maintained, is here reflected in the division which itself yields the ethical. Kant, for example, in wanting to ground the possibility of ethics in a universal conception of duty, had to leave unposed the question of the basis of duty itself. There is no obvious answer to this question, or at least not one that avoids the potential circularity that would result from an attempt to resolve the difficulty by resorting to an argument based on the need for such a conception of duty.

The specific importance of the *Discours de la Méthode* is that there Descartes tries to establish a methodology which both necessitates, and depends upon, forging a radical distinction between the domains of error and truth. The articulation of this position involves the use of the language of ethics, especially in framing the relationship foreseen between the text and reader. The major question addressed by the commentary is how successful Descartes has been.

2.2 *The critique of common sense*

Early in what proved to be a prolific correspondence, Descartes wrote a letter to Mersenne in which one of the principal subjects of discussion was the aesthetic.[5] The problem posed by the aesthetic was that it did not lend itself to scientific study. Not only was pleasure to be divorced from truth, it could not even be spoken of in the same breath. Pleasure was to be relegated to the domain of opinion, imagination and the non-universalisable. Pleasure in relation to the aesthetic lends itself to diversity, to what are called in the *Discours*[6] the 'diversity of our opinions', which stand in opposition to a method which 'teaches us to follow the true order'. The clash is between the multiplicity of opinion and the unitary nature of truth. The multiplicity of opinion cannot be allowed to impinge upon the singularity of truth. The distinction between multiplicity and unity is not imported into the *Discours,* rather it flows within it. Descartes in Part Two suggests that the advantage of method lay in the fact that, 'there being only one truth about each thing whoever discovers it knows as much about the thing as can be known'. The distinction between the one and the many is not unique to Descartes for it is precisely in these terms that Plato adumbrates his own epistemology in the *Republic.* Diversity was and is the domain of error, sin and temporality. The inclusion of temporality is of course a consequence of Platonism insofar as things (*ta onta*) are temporal while the forms are eternal.

The dream of philosophy is, as has been mentioned, to overcome the deception stemming from the interplay of diversity, sin and temporality and thus to lay a firm foundation in terms of which truth, in Descartes' sense, and morality, in Kant's can be guaranteed.[7] In the case of Descartes, the method which consists in following 'the true order' is that which 'contains everything that gives certainty to the rules of arithmetic'. The question that must be asked is, how successful has this been? This is not to ask whether or not philosophy thinks itself to have dismissed or excluded the fictive. Rather, philosophy must be taken at its word, and the cost of the dismissive gesture must be evaluated in terms of its consequences for that particular conception of philosophy.

As a preliminary step toward an evaluation of Descartes' autobiographical attempt to give philosophy a foundation in method, it is essential to trace his critique of common sense. It is from this critique that emerges the importance and necessity of a methodological basis for certainty, because fundamental to Descartes' strategy is the recognition that common sense cannot of itself yield certainty. In order to trace the critique I shall examine in some detail a number of passages from the *Discours* which will pave the way toward an understanding of a fundamental moment in the text, namely Descartes' estimation and evaluation of his own intellectual training. This will create a position from which to assess what it is that the *Discours* is attempting and therefore how it should be understood.

The first passage to be examined is the conclusion of Part One of the *Discours,* which can be viewed as a justification by Descartes of his approach.

> It is true that while I only considered the customs (*les moeurs*) of other men, I hardly found anything to assure myself and I noticed almost as much diversity as I had previously among the opinions of the philosophers. The greatest profit that I derived from them was that, seeing several things which, even though they seemed to us extravagant and ridiculous, were still commonly accepted and approved by other great nations, I learnt not to believe anything too firmly of which I had only been persuaded by example and by custom, and thus I gradually freed myself from many errors which can obscure our natural light and render us less capable of responding to reason.[8]

This important passage begins with the assertion that mere empirical observation cannot provide certainty. Descartes cites two reasons why this is the case. The first is that observation on its own is ill-informed and not of itself systematic. This is a position advanced by Descartes throughout his philosophical writings; namely, that while all knowledge is based on observation and is thus empirical in nature, error is also based in observation. Observations therefore must be ordered and dealt with systematically in order to avoid the possibility of deception. The second reason is that the 'customs of other men' are diverse.

Diversity may and probably will mean that various groups of people will hold conflicting attitudes in relation to the same activity or thing. Clearly, if truth is unitary in nature, then diversity strikes at the heart of truth itself. Consequently, Descartes argues, a thing cannot be true simply in respect of its being believed to be true. Belief, even if held by a majority, does not provide a thing with its truth. As Descartes states, the 'extravagant and ridiculous' are accepted by some as true. The consequence of this position was Descartes' decision to believe no longer those things where the basis of the belief was either custom or example. As he realised, to abandon such beliefs was in fact to abandon the world, for not only was the world the source of error, it also restricted the operation of the understanding. Descartes is not suggesting that knowledge is or ought to be *a priori*, but simply that knowledge of the world is only possible if method both instructs and guides claims about the world. The universality of method would yield unitary truths and thereby would give knowledge of the world a firm and secure foundation.

The critique of custom is joined to an intriguing use of the image of the one and the many in Part Two of the *Discours*:

> . . . it is custom and example that persuade us rather than any certain knowledge; nevertheless the plurality of voices is worthless as a proof for truths that are slightly difficult to discover, thus it is much more likely that a man on his own will find them than an entire people.[9]

This passage is of great strategic importance in the *Discours*, for it provides part of the foundation for the autobiographical nature of Descartes' project. If one person is more likely than the majority to discover difficult truths, then there is no reason why Descartes should believe anyone apart from himself. The autobiographical component, while important, should not in this passage be overestimated, for it is conditioned by and thus emerges from the distinction between the one and the many. In other words, the emergence of the importance of the solitary voice is part of the strategy involved in overcoming the possibility of deception, in so far as error is related to the many and truth to the one. This passage need not be interpreted therefore as an argument for the centrality of the *cogito* in the determination or discovery of truth, for it indicates that a more complex manoeuvre is at work.

Descartes has once again dismissed habit and example as themselves being able to furnish the basis of truth and goes on to suggest that the reason why 'popular beliefs' are held is best accounted for in terms of habit and example. The important aspect of error is that it is linked to 'popular belief' and sometimes has its origin in such beliefs. The consequence of this link is that the overcoming of error must involve a two-fold strategy. The first component of this strategy would be to give up 'popular beliefs', while the second would involve a search for a method in terms of which those popular beliefs that are true can be shown to be true. Descartes wants in the end to be able to distinguish between a set of beliefs that are only believed because of habit and custom, and a set of beliefs the certainty of which has been established, and are thus not held to be true in respect of being believed, but are held to be true because truth has been given a transcendental guarantee. In Descartes' words a 'man' on his own is more likely to overcome the impetus of popular assent than a crowd. Knowledge is not phronesis but legislation. Method and the knowing subject act in unison but the method that establishes certainty pre-exists and post-exists any one instance of its use and is therefore that which allows the legislator to act. The certainty of the *cogito* is grounded in the certainty of method. The problem of the foundation of method itself remains unresolved.

The final passage on which I want to comment concerns a description of error that is advanced in the First Part of the *Discours*:

> However, it may be that I am wrong, and it may be only a bit of copper and glass that I take for gold and diamonds.[10]

Descartes' claim is that we are often mistaken about that which is at hand or which concern us, in other words by things '*qui nous touche*': the important point is why this is so. On what is this particular form of error based?

The possibility of a relationship, let alone a confusion, between copper and glass on the one hand and diamonds and gold on the other, could only be provided by resemblance. 'I' may be mistaken only because one resembles the other. The imagination moves from one to the other along a chain of resemblances. The theme of resemblance is well known in Descartes' philosophical work. It is continually cited as a cause of error and is often invoked in explanations of how the imagination works. There are however two aspects of resemblance that need to be emphasised here. The first resemblance can be described as a chain of interconnection where metonymy provides the possibility of movement along the chain, then this movement has the consequence of robbing the thing of its specific identity. The reason why metonymy accounts for this movement is that resemblance differs from representation in that the connection between two things is established by an element in each. The possibility of substitution belongs to shared elements rather than to the thing iself. Clear and distinct perception can be understood, therefore, as an intervention into a sequential chain which breaks it into separate identifiable links and thus re-establishes the identity of each specific thing. The links re-established as discrete identities means that they have acquired the self-identity envisaged in the definition of thing (*la chose*) in *La Logique du Port-Royal*. 'I call a thing what one can conceive as subsisting by itself and as the subject of all that can be conceived of there'.[11] The thing existing in-itself can be represented. The importance of representation is outlined in the *Second Responses* where Descartes argues that the knowing subject reprsents the world to itself in the form of an idea. If it were not the case that clear and distinct perception overcame resemblance, knowledge of the world would be impossible, since the co-extensivity between sign and thing that characterises Cartesian representation would itself be impossible. Knowledge therefore is premised on the possibility of overcoming not simply the threat of the imagination, but more importantly the interrelationship between resemblance and metonymy, in other words the basis of fiction itself.

The second aspect on which I want to comment concerns elements of the historicity of resemblance. The need to overcome resemblance is a constant theme in the history of philosophy. St. Augustine in *De Trinitate*[12] suggests that one of the ways in which 'mind' (*mens*) goes astray in its attempt to respond to the demand 'know theyself' was to think that it resembled the contents of mind and thus conceived of itself as a physical object.

The problem of resemblance and therefore of error, as Augustine indicates, cannot be divorced from the general question of mimesis and thus of fiction. Mimesis or imitation is itself inextricably linked to, if not originating in, the distinction drawn by Plato between the 'idea' and the physical instantiation of the 'idea'. In the *Republic*, in his treatment of the activity of the cabinet-maker Plato draws a distinction between the Form (*to eidos*) of the bed and a particular bed, and goes on to say that the cabinet maker,

> . . . does not make what *is*, the product is not that which is, rather it is like what is without *being* it. (Section 597a lines 3–5.)

The distinction between 'what is' and that which is similar to 'what is' is not restricted to the category of physical object and the 'idea' (*to eidos*) of the particular object, as it is this distinction which makes problematic one particular mimetic activity, namely poetry.

In sections 598–601c of the *Republic* Plato's preoccupations centre around his claim that the poet (here Homer) 'has no hold on truth'; (*tes aletheias oukh aptesthai*, Section 600e line 7). The inability of the poet to control the truth of what is said will allow the 'poetic' interchangeability of glass and diamond, for this does not necessitate knowledge of either. The other important aspect is that the initial likeness opens up the possibility that glass could be believed to be diamond. It is the possibility of being deceived about the exact nature of material objects that Plato is in general attempting to point out, though in this instance he is making the further claim that poets make use of the way in which there is the potentiality for deception in their creation of poetry. While Plato does not express it thus, it can nonetheless be argued that not only does poetry involve mimesis, it also makes use of, if not depends for its poetic quality, on metonymy. The consequence is that overcoming deception will involve a necessary reciprocity between the specific ontology of the object of knowledge (namely an ontology of stasis) and knowledge itself. Plato establishes this connection in the *Republic* Sections 477a–b, where he states the following:

> . . . knowledge is related to what is, and knows what is as it is. (Section 477b, lines 8–9).

This does of course commit Plato to the existence of a fundamental reality of things both temporal and eternal about which deception is possible. It is the retention by Descartes of this ontological structure that situates the concerns of Cartesian thought within the ontologico-epistemological framework that commences with Plato. The relationship between truth and fiction must be always envisaged as taking place within this framework.

For Descartes to think that glass and copper were diamond and gold would be to engage in a fiction that was only possible because of resemblance, though this could be overcome by that particular form of cognition which would secure and ensure genuine knowledge of the world. Related to this is the fact that the threat of resemblance and imitation concerns the domain of 'common sense', and that therefore the guide of method is needed to steer knowledge of the world away from resemblances and towards things themselves. The existence of deception entails that the search for the foundations of knowledge is imperative.

The critique of 'common sense' involves, therefore, the importance of method; the knowing subject's subordination to method; the recognition that custom and example cannot provide the basis of truth; the necessity of the destruction of the chain of resemblances by clear and distinct perception (for this alone will facilitate the form of perception central to truth, namely of seeing one thing apart from the other);[13] and finally, the triumph by method over the continual threat of fiction. These themes are also at play in Descartes' evaluation of his own school life.

2.3 *The fable of overcoming fables*

Descartes' assessment of the training he received at *Le College des Jesuits de La Fleche* forms part of his general argument in favour of abandoning old educational procedures. The argument is in part based on his recognition that certainty and truth could not be based on anything that he was taught. He did suggest that some value could be gleaned from his lessons. 'I knew that the languages learned there are necessary in order to understand the books of the ancients; that the charm of fables awakens the mind'.[14] In the end he decided, however, that such a diverse range of subjects and the knowledge to be gained from them were unnecessary for the task at hand. 'But I thought I had already given enough time to languages and also to reading the books of the ancients, and their histories (*histoires*) and their fables (*fables*)'.[15] Not only were such subjects diversions, in the case of fables there was a greater problem; 'fables make us imagine many events as possible which are not'.[16] In other words, fables introduced not just the threat of fiction and therefore deception, but its actual presence. The methodological was intended to triumph over the fabulatory. The instantiation of method was understood therefore as at the same time announcing the demise of fiction, in so far as it was a threat to truth, while opening an untransversable gulf between the possibility of error and the possibility of truth. Error's link to the imagination and thus to the fictive was to be permanently sealed off from the interrelationship between understanding and knowledge. The consequence would be that the actual existence of error could no longer threaten the presence of truth, and also that which brought it about, namely method.

Having sketched some of the prerequisites for a discourse on method as well as having indicated some of the elements presupposed in such a discourse, it is essential to examine Descartes' own understanding of the *Discours de la Méthode*. This is not a recourse to intentionality and thus to viewing the *Discours* as a mere consequence of intention, rather it is to respond to the autobiographical component of it in which it is argued, as we have already noted, that the solitary individual is more likely to discover complex truths than the multitude. If Descartes is able to claim, as a consequence of the critique of common sense, that the individual rather than the crowd is more likely to discover the truth, then it seems reasonable to take this conception of the *Discours* as itself presenting the frame within which the *Discours* may be interpreted.

In order that this be achieved we need to return to Part One of the *Discours* in which Descartes advances three different though interrelated conceptions of what the text actually is. In each case there are additional factors which make these comparisons of special interest. The first comparison is the following:

> . . . I shall be very pleased to show in this discourse the ways that I have followed and to represent my life in it as in a painting (*un tableau*), so that everyone may judge it and that, learning from public response the opinions they have of it, this will be a new means (*moyen*) of instructing myself that will add to those which I am accustomed to using.[17]

According to Descartes the *Discours* will have the same status as a painting (*un tableau*) and in addition not only will it be judged by the many, the author now sees himself as constrained to respond to the many. The crowd's response will serve the author as a means of self-instruction, thus almost inverting the way in which the relationship between the one and the many had previously functioned in relation to the search for truth. This comparison and the related inversion will have to be explained in detail. However, prior to setting out on this task, it will be best to set in motion the other two comparisons made by Descartes, both of which occur in the same passage:

> (but) only proposing this work as a history (*une histoire*) or if you prefer as a fable (*une fable*) in which among other examples that can be imitated (*on peut imiter*) you will perhaps also find several that it would be correct not to follow.[18]

In this passage Descartes' claim is that the *Discours* is proposed not only as a story or history (*une histoire*) or a fable (*une fable*). However, implicit in this claim is the further point that the reader's response is delimited by imitation, as all possible responses centre around the question of whether or not to imitate the story being told. In other words, the writer of an autobiography is suggesting that after having read his life the reader may choose to live (or not) in the 'same' way. The problem of the 'same' is of course vast. Indeed it is this 'same' that is at the heart of Plato's final condemnation of the practice of imitation in the *Republic* (Section 602a–b). Even though we shall return to the problem of the 'same' it is worth noting that part of Plato's objection is simply that imitation is possible independently of any real knowledge of what is is that is being imitated. It is after all the 'same' that plays into the hands of deception via resemblance.

The first problem concerns examining those things which are at play in the comparison of the *Discours* to a painting. The complicating factor is that Descartes is interested in the public's response to this specific painting. It is to another text, the first draft of which had been written at the time of his writing the *Discours,* that we can turn to see what is at play in the comparison between text and *tableau*. In the first *Meditation,* one of Descartes' initial moves is to prepare the way for the argument that the possibility of deception need not stand in the way of attaining truth and certainty. Part of these preparatory procedures involves establishing a distinction between the idea of thought, and the content of the idea or thought. The result of this is to provide Descartes with the conceptual manoeuvrability to overcome the threat posed to truth by dreaming. Establishing this distinction involves Descartes in a detailed consideration of painters whose paintings contain imaginary events or animals.

> At the same time it must at least be acknowledged that things which are represented to us in sleep are like the tableaux of painters which can only have been formed from the resemblance of something real and true (*de réel et de véritable*) and it is thus for at least these general things, namely eyes, heads, hands and all the rest of the body, are not imaginary but true and existent. For painters even during the careful construction of Sirens and Satyrs by bizarre and extraordinary forms cannot attribute to them entirely new natures but only make a certain mixture or composition of the limbs of various animals; or if their imagination is extravagant enough in order to invent something so new that the same thing had never been seen before and thus their work represented a thing purely fictitious and absolutely false (*une chose purement feinte et absolument fausse*) then at least the colours of which they are composed must be real (*véritables*).[19]

In this important passage Descartes' first move is to compare the reality of the contents of dreams with that of paintings whose reality is completely dependent upon their content resembling something that is real and true. The existence of a non-fictional element in both dreams and paintings is based on there being a resemblance between all or some of their content and something outside of the dream or painting whose existence could be assured. As the passage indicates, resemblance is enacted via representation, for resemblance furnishes paintings with their reality to the extent that what is represented resembles that of which it is a representation. Descartes foresees that the immediate objection to this characterisation would be to argue that some paintings have contents that are so fabulous, there could be no possibility of a resemblance between what is represented in the painting and something 'réel', let alone 'véritable'. This argument is summed up in the question of how something can be represented in a painting when there is no apparent resemblance between the fabulous content and a reality beyond the *tableau*. It is essential to notice here that this objection depends upon a necessary and functional reciprocity between resemblance and representation.

Descartes responds to this already foreseen objection with the claim that even within fictions there are elements that are not fictional, that is, elements which both resemble and represent something '*réel*'. Even in the extreme case where the painting presents 'a thing purely fictitious and absolutely false', there is still an element that is true, for there remains a resemblance between, for example, the colours in the painting and the colours themselves.

Without having to dwell on every aspect of Descartes' argument it is clear that one of the inevitable consequences of the attempt to divorce fiction from reality is that the reality of the painting (or the representation) can no longer be said to lie in the representation *tout court* but only in the relationship between the elements and what they resemble. The difficulty is of course that even if the *tableau* were not fictional and there appeared to be a resemblance between the *tableau* and the *réel*, this could never be established with certainty so long as the reality of the painting depended upon resemblance. It is, after all, precisely resemblance that is at the heart of error, and thus it is that which needs to be replaced by clear and distinct perception. The problem is compounded by the fact that the reality of the painting does not lie exclusively in the painting but only in the elements of the representation. The truth of the *tableau* can be neither pinned down nor guaranteed for it is at every moment neither completely false nor completely true. The *tableau* is the exemplar of fiction.

The two questions that arise from this conclusion are, firstly, whether or not the *Discours de la Méthode* is a fiction, and the related question of what would secure the truth of an autobiography? The answers revolve around the possibility that the text truly represents the life as it really happened. Any other configuration would entail a displacement between life and representation with the consequence that the representation no longer represented, and as such the *Discours* would indeed be a fiction.

There needs to be a coextensivity between the representation and what is represented. It is not surprising that the definition of 'sign' advanced in *La Logique du Port-Royal* is couched in terms of this coextensivity. That the sign is a metaphor for the way in which the *Discours* can be true and not a fiction, can be seen in the definition of sign; 'the sign encloses two ideas: one of the thing which represents: the other of the represented thing'.[20] From the definition it is clear that to the extent that the text remains a sign, it eludes fiction. The dilemma is that the conditions that the *Discours* must fulfil to be a sign are exactly those that it cannot. While there is an interdependence between representation and resemblance in so far as the relationship between the 'the thing which represents' and 'the represented thing' cannot be based on resemblance, representation must involve more than resemblance. The structure of the sign demands coextensivity, and therefore the necessity to exclude resemblance is due to its connection with metonymy. The result of the connection between them is the blurring of the identity of those things within a metonymic chain and it is the aim of representation to represent the actual identity of 'the represented thing'. In this instance therefore, resemblance undermines the possibility of the *Discours* remaining a sign because the necessity of its presence undermines the coextensivity inherent in the structure of the sign. In other words, the necessary presence of resemblance means that the structure of the sign is never secure. The result of the need to exclude resemblance is that the *Discours* is irreparably divided from the possibility of its own truth. What would make it true is what removes it from the position where it could attain its own truth.

The other important consequence is that the method which is represented in the *tableau* and whose province is the discovery of truth is robbed of its certainty, not just by the fiction of the *tableau*, but also by the presence of the method in the *tableau*. The reality of the painting is provided by resemblance, and therefore the painting cannot both contain and present the truth. What is true in the *tableau* is true for reasons that are found beyond the frame, namely in the resemblance between the representation and the real. The *tableau* is literally unable to contain the truth.

It is, however, only in terms of the *Discours* understood as *tableau* that it is possible to understand the way that Descartes expects his readers to respond to the text. The peculiarity of the envisaged response is that it stands in contradistinction to his initial formulation of the relationship between the one and the many. The initial formulation was based on the fact that because people can and do hold false beliefs, it is more likely that an individual working on his or her own will discover the truth. The consequence of this is that the untutored response of the many should count for little in the search for truth. In the passage under consideration Descartes states that his response to public judgement will amount to another 'means' (*moyen*) by which he can instruct himself. Public opinion is on the one hand diverse, not grounded in method, not advanced from a position that ensures accurate judgement, and is thus that which should be excluded from the domain of truth. On the other hand, however, the response of the many is to be taken at face value as itself comprising a legitimate response, and moreover a response that is instructive in the general search for truth. It must be remembered that the second response is to a *tableau* and therefore it is the *tableau* that introduces the paradoxical position of public opinion. Indeed, this position is only possible if the initial structure of the one and the many is inverted. The reason that the inversion has this result is based on the function and necessity of resemblance in the contents as well as in the understanding of the *tableau*. The necessity of resemblance in order that the *tableau* be a *tableau*, coupled with the incommensurability between a unitary conception of truth and resemblance, combine to result in the abnegation of unity. The inversion dislodges the centrality of a unitary truth in so far as this conception of truth is structurally similar to the sign in *La Logique du Port-Royal*. It also dislodges the centrality of the solitary individual in pursuit of truth. In this regard, the inversion has the effect of casting doubt on both the efficacy as well as the centrality of the subject, that is, the 'I' of the *cogito ergo sum*.

As the inversion means that potential diversity is no longer excluded, the possibility of error is reincluded, and therefore the strategy that was intended to divorce truth from the possibility of error has resulted in uniting them. The comparison between text and *tableau* has the consequence of dividing the text, turning it into a paradox that could only ever be overcome by suppresing those moments within the text that gave rise to the division, and hence the paradox. It is thus that the unity of the text is contingent upon the suppression of the logic of paradox.

Not only did Descartes suggest a comparison between the *Discours* and a *tableau,* he went further and suggested that it could be understood as *une histoire* (a story or a history) or *une fable* (a fable). I will not dwell on the suggestion that the *Discours* can be read as *une histoire,* but only point out that Descartes' subsequent criticisms of *l'histoire* are usually seen as an attempt to set the limits of Virgil's *Aeneid;* in other words, to set the limits of fiction. It is perhaps sufficient to note that the term *l'histoire* in the context of its appearance in the *Discours* implicates the text, yet again, within philosophy's constant struggle to overcome fiction and to do more than merely tell stories.

Of more direct importance are the elements at play in reading the *Discours* as a fable. The first problem is how we are to understand the fable. While it is important to realise that the moral of a fable is practical in nature, it should be remembered that the fable can never present an obligation, that is, it cannot present anyone with a duty to act in one particular way rather that in another. In part this forms the basis of Rousseau's objection to the fable. The reading of fables to children, he argued, could result in their being as easily attracted to vice as to virtue. The fable does therefore lack the force of an imperative. This absence will emerge of as being of great importance; however at this stage we must remain with the question of the relationship between the fable and philosophy. La Fontaine in the preface to his *Fables* wrote of the connection between philosophy and literature in the presentation of his version of the death of Socrates, as recounted in the *Phaedo.*

> Socrates had been condemned to death but his execution had been postponed because of holidays. Cebes went to see him on the day of his death. Socrates said to him that the Gods had advised him several times during his sleep to apply himself to music before he died. First of all he had not understood what the dream meant: for, as music did not make men better what good could be derived from such an application? There must have been some mystery involved especially as the Gods had not tired of sending him the same message. . . . He had noticed that there was music and poetry without harmony: but there was none without fiction; and Socrates only knew how to tell *the truth.* Finally he found a solution which was to choose fables that contained something of the truth, such as those of Aesop. He used therefore the last moments of his life putting them into verse.[21]

If, as La Fontaine suggests, a fable contains something of the truth, then its success in conveying it must involve fiction; for there is no good poetry 'without fiction'. The consequence of the link between poetry and fiction is that poetry and fiction depend for their communicative power on *persuasio*. In other words it is *persuasio* rather than *scientia* that is fundamental to the way in which fables convey truth. The difficulty is of course that while *scientia* is the mode of knowledge proper to method, and thus that which the *Discours* must impart, it is exactly this mode which is excluded in order that the fable can function effectively as the conveyer of truth. A fable depends upon persuasion rather than on the recognition of truths which are the consequence of clear and distinct perception.

This, however, is not the whole of the matter, for Descartes has also said that the reader may respond to the fable by imitating (or not) its content. The aspects that readers find of value may be imitated while others can be ignored. There are many areas of further discussion opened by imitation, but the most important concerns the relationship between imitation and the failure of a fable to present an imperative.

It has already been noted that fundamental to the critique of common sense is the position that resemblance cannot provide the basis of knowledge. Knowledge involves knowing the thing as it is, and it is to that extent that the object of knowledge is unitary in nature. Linked to this conception of knowledge is what could be called an epistemological imperative, in so far as because neither truth nor knowledge admit of degrees, the true as a consequence has to be recognised as true. This point is further reinforced by the fact that as the 'natural light' is universal, the truths which are the consequences of the application of method and recognised as such by the 'natural light' are also universal. Universal truths have to involve an epistemological imperative.

An important aspect of such an imperative is that it be recognised as an imperative. Responding to it must involve, therefore, the full recognition both of the imperative *qua* imperative as well as those elements essential to establishing the imperative *qua* imperative. Two of the most important of these other elements are the recognition of the failure of resemblance to ground knowledge and the necessary overcoming of the threat of fiction. Both are essential in order that the epistemological imperative be recognised.

The problem is obvious. This *Discours* is not just described as a fable by Descartes, it has already been noted that it is constrained to adopt the figure of fiction. It is not, therefore, that there is an incompatibility between the claim made by Descartes that the *Discours* can be read as a fable and the need to establish a discourse on method. If it were, then the only difficulty would be that the text contained a contradiction. The diffficulty is more complex because the text, in order that it be a methodological treatise, has to adopt a fictional mode of presentation. The problem, therefore, is not contradiction, but paradox. It is the paradoxical element that emerges in relation to the imperative. A fable, which the fictionalised *Discours* has become, cannot advance that which it must to be a discourse on method, namely an epistemological imperative. The *Discours* has become the site of the paradox stemming from philosophy's attempt to rid itself of fiction.

As a fable cannot present its readers with imperatives and as the *Discours,* in order to be a methodological text, must be followed to the letter, its only possibility of surviving as the presenter of methodological imperatives is by the suppression of the fictive within it. It is this question, of suppression and therefore of reading suppression, that will serve as a conclusion.

The reason why this kind of concluding commment is needed is simply that the *Discours* is seldom read as a text which is in the end forced to relate philosophy's fabulous end. Ths question emerges, therefore, of the relationship between a conventional reading of Descartes' text, which reads it *à la lettre,* and therefore as a discourse on method, and the reading deployed in the preceeding pages.[22] This is by no means an easy question to answer, and all that will be attempted is to offer an indication of a fuller response.

The major point of divergence concerns the way in which the emergence of a philosophical argument or position is interpreted or understood. The difference centres on the importance attributed to textuality. The conventional reading seems in some way to rely on what it takes to be a non-textual outside that provides the means or reasons for interpreting the text. The outside could be context, or perhaps even an investigation of the viability of the arguments at play in the text under consideration. In each instance it is the purportedly non-textual that provides access to the text. The proof that the outside is thought to be of a fundamentally different nature to the textual inside it that it is not itself subjected to any form of analysis. It is as though the use of contexts or the investigation of arguments were themselves beyond interpretation. The claim being made is not that contexts are beyond interpretation, but rather that the process of having recourse to them is. The consequence of subjecting both contexts and the use made of them to interpretation, and thus of emphasising their textuality, is not that philosophy becomes ahistorical and no longer interested in historically delimited intertextual studies; it is rather that the claims of history and historicity lose their automatic privilege, and the conditions that govern the emergence of the text become of central importance. History is reinscribed within the practice of interpretation and is no longer in the untenable position of attempting to remain outside, governing interpretation. The position of government moves from the outside to the inside, and in so doing dissolves the inside/outside opposition. The other consequence is that history becomes textual. It is more than simply concerned with texts, for it is itself a textual event.

The emphasis on textuality means that the text cannot be given or attributed an organising principle from outside the text itself. The direct result is that the text is no longer a simple unity but rather is the site where the interplay of the conditions governing its existence takes place. There can be therefore no general statement concerning the conditions at play in all texts. Once again, this would be to erect a governing theory which sought to establish an outside that would control the inside. The absence of a general theory of texts (as opposed to a description of texts) means that what is at issue is a reading of a particular text. In this instance the text in question was Descartes' *Discours de la Méthode* and its attempt to forge a radical distinction between truth and fiction. In focusing on the unfolding of the text, what emerged was the impossibility of this task. It is at this stage possible to offer an interim conclusion, namely that the *Discours* is indeed a fable, one recounting philosophy's oldest aspiration. The moral of the fable is practical, in so far as it suggests, though no more than suggests, a way of reading texts.[23]

Notes

1 As the major concern of this paper is Descartes' *Discours de la Méthode* I have not attempted to provide a justification for this reworking of philosophy's goal. I have however provided some justification in my 'Wisdom and science in Augustine and Descartes', *Ideas and Production,* 2, 1984.

2 This paper is collected in M. Heidegger, *Wegmarken,* Frankfurt am

Main, 1967. English translation in, *Martin Heidegger: Basic Writings,* ed. David Farrell Krell, London and Henley, 1978.

3 This paper is also collected in Heidegger, *Wegmachen,* see in particular pp. 23–5.

4 The passage I have in mind is the following and it does I think show the importance Heidegger attaches to the 'oversight'. The translation comes from the English edition by David Farrell Krell, *Nietzsche, Vol. 1: The Will to Power as Art,* London, 1981, p. 149.

That the question of the essence of truth is missing in Nietzsche's thought is an oversight unlike any other; it cannot be blamed on him alone, or him first of all – if it can be blamed on anyone. The 'oversight' pervades the entire history of Occidental philosophy since Plato and Aristotle.

Clearly more needs to be said about Heidegger's own position if only because of his special understanding of the term 'essence'. On this latter problem see George A. Ghanotakis, 'Unscrambling Heidegger's notion of "essence": a consideration of some topographical and thematic difficulties', *Journal of the British Society for Phenomenology,* 15, 1984.

5 Descartes, *Oeuvres Philosophiques,* ed. F. Alquié, Vol. 1, Paris, 1963–7, p. 251. This letter is also discussed by Peter France in *Rhetoric and Truth in France,* Oxford, 1972, p. 41. France's analysis of Descartes follows a different path to the one pursued here.

6 All references to the *Discours de la Méthode* are to the edition established by Étienne Gilson, Paris, 1967.

7 For a discussion of this point that follows a similar though not the same line, see Richard Rorty, *Philosophy and the Mirror of Nature,* Oxford, 1980.

8 Gilson, ed., *Discours,* p. 10. All translations are my own.

9 *Ibid.,* p. 16.

10 *Ibid.,* p. 3.

11 A. Arnauld and P. Nicole, *La Logique ou L'art de Penser,* Paris, 1970, p. 73. The value of the *Logique* is that its Cartesian nature and the fact that it was continually modified between 1660 and 1683 means that it provides a contemporary commentary on Descartes' own writings.

12 For a further discussion of this point see my paper mentioned in footnote 1 above.

13 This point is made by Descartes in *La Recherche de la Vérité* in *Oeuvres Philosophiques,* ed. Alquié, Vol. 2, p. 1140.

14 Gilson, ed., *Discours,* p. 5.

15 *Ibid.,* p. 6.

16 *Ibid.,* pp. 6–7.

17 *Ibid.,* pp. 3–4.

18 *Ibid.,* p. 4.

19 The text of the *Meditations* that I have used is *Meditationes de Prima Philosophia. Méditations Métaphysiques,* Texte latin et traduction du Duc

de Lyons. Introduction et Notes par Geneviéve Rodis-Lewis, Paris, 1966, pp. 20–21. The virtue of this edition is that it provides both the Latin and the French text on opposite pages. The important claim of the representation of something both fictitious and false is borne out in the French, quoted in the text; and in the Latin 'ficitium sit & falsum'.

20 Arnauld and Nicole, *La Logique,* p. 80.

21 La Fontaine, *Fables*, Paris, 1966, pp. 27–8.

22 I do not wish to pretend for a moment that this paper is a complete departure from all other interpretations. Indeed, it is indebted, as are all current reinterpretations of Descartes, to Jean-Luc Nancy's important collection, *Ego Sum,* Paris, 1979. Another important recent interpretation of Descartes is that by Sylvie Romanowski, *L'Illusion chez Descartes,* Paris, 1974. The question of imitation receives an interesting discussion in John D. Lyons, 'Subjectivity and imitation in the *Discours de la Méthode*', *Neophilologus,* 66, 1982.

23 I wish to acknowledge the help given by John Christie and Sarah Richmond who provided extremely valuable comments on an early draft of this paper.

3 *George MacDonald Ross*

Hobbes's two theories of meaning

3.1 *Introduction*

It is always tempting to look out for anticipations of modern ideas in the writings of past philosophers; and it is exciting to discover an anticipation that never seems to have been noticed before. However, this attitude betrays a concern for priority which can lead to serious distortions in writing the history of philosophy.

In certain cases (and such cases are the exception rather than the rule), it is indeed quite proper to emphasise a philosopher's priority in evolving a new idea or theory. When philosophers have manifestly succeeded in transforming the terms in which problems are discussed, then the novelty of their thought is a significant historical fact about them. It will be evident through their having given rise to new philosophical schools, or at least through their having become regular objects of adverse criticism.

Even here, however, there are difficulties. When philosophers have attracted a following, the followers have a natural temptation to ascribe their own ideas and interpretations to the founders themselves. But these followers often do themselves an injustice by discounting their own contributions in sharpening up and embellishing the ideas which formed the starting-point of their endeavours. Even the most original thinkers have to struggle in order to escape completely from the concepts and theories that form their intellectual milieu, and against which they are reacting. They are not in any specially privileged position for predicting how their ideas will evolve in the hands of their followers, nor is it necessarily the case that they would approve if they knew. This is why there is no paradox in saying, for example, that Kant was not a Kantian, that Marx was not a Marxist, or that Wittgenstein was not a Wittgensteinian. Seminal thinkers are often more old-fashioned than their disciples would like to think, and yet in other respects more original than they recognise.

The situation is far more problematic in cases where philosophers are alleged to have made *isolated* discoveries not taken up by their contemporaries or immediate successors. It could even be held that thinkers can never be significantly out of their own time, on the grounds that the language they used to express their novel ideas would necessarily be a private one, and would therefore lack any objective criteria for meaningfulness. Someone like Nietzsche might be held up as an example of a philosopher who lapsed into unintelligibility as a result of his efforts to transcend the concepts of his public. I have two reasons for rejecting this extreme position. The first is that I do not accept the extension of Wittgenstein's 'private-language argument' beyond the context within which it arose, viz. the naming of private mental states. I shall touch on this point later in connection with Hobbes, in Section 3.3.3 below. My second reason is that no conceptual change is possible unless people can in fact give expression to ideas unique to themselves. The historian is sometimes in a better position to recognise these ideas for what they are than the original thinker's contemporaries, if language has subsequently evolved in such a way as to make those ideas more readily expressible. On the other hand, even if we accept the *possibility* of thinkers being ahead of their time, there are almost intractable difficulties over establishing any hard-and-fast criteria for deciding when they are in fact thinking originally; when they are merely formulating relatively conventional ideas, if perhaps in an unusual way; and when they are simply writing nonsense.

The problem is essentially one of *interpretation*. The historian of philosophy is more likely to be conscious of the problem when translating from a foreign language, especially a dead one such as Latin. In these cases it takes a certain boldness to give an interpretation which imputes senses to words which cannot be supported in terms of established usage. The problem is more insidious when the philosopher in question wrote in the same language as the commentator, since the meanings of words often change, and it is all too easy to read the text anachronistically. There can be *faux amis* within a single language at different times, as easily as between different languages at the same time.

Suppose an interpretation has been offered of a historical text, which makes it appear original and modern. One good criterion for accepting the interpretation would be that no other interpretation is possible; but a negative like this is always hard to establish. Perhaps the interpreter has insufficient information about contemporary usage; or perhaps the author had not properly thought through what he was writing. In most cases, a little ingenuity will provide an alternative interpretation according to which the author is saying something much more bound to the concepts and beliefs of the time. If so, we obviously ought to accept the interpretation which is least strained, and requires the least ingenuity to make it plausible. But this is again difficult to establish, especially without begging the question of what the author of the text is *likely* to have believed. An interpretation might *seem* strained, because it requires a considerable amount of background explanation – but such an explanation could be fully justified as providing the context which would have informed the intuitive understanding of a contemporary, and which we now lack.

In general, it would seem reasonable to adopt the principle that a conservative interpretation is more likely to be correct than a radical one – in other words, writers are more likely to mean what their immediate readership would naturally take them as meaning, rather than what readers in a future age might ascribe to them. A radical interpretation would require supporting evidence, such as its coherence with the rest of the philosopher's system, or explicit comments by the philosopher's contemporaries (even puzzlement might constitute evidence that certain thoughts transcended contemporary concepts).

In addition to the difficulties involved in establishing that modern ideas have been prefigured in a historical text, the search itself distorts the historical process. At best we might discover that earlier thinkers were not quite as unsophisticated as they may have appeared when seen through the concepts of the last generation of commentators. But if we set out to find anticipations of our own ideas, it is highly unlikely that we will gain anything by discovering them. We will see only what we already know, and remain blind to ideas which might constitute a genuine and valuable alternative to some of today's assumptions.

In this paper, I shall be examining a particular case of a philosopher, namely Thomas Hobbes, whose ideas on language seem to prefigure certain modern theories, and to do so in a way which was not understood or taken up by his contemporaries. Moreover, the question of how to interpret Hobbes's texts, both Latin and English, raises precisely the range of issues I have just outlined.

The *locus classicus* of Hobbes's account of language is *De corpore*, Part I, ch. 2. I shall refer to the theory expounded in this chapter as Hobbes's 'official' theory of meaning. Hobbes has suffered from relative neglect as a theoretician of language, but enough has been written for a 'traditional' interpretation to be identified, according to which Hobbes is, in effect, a predecessor of the positivists and the early Wittgenstein. The ultimate purpose of this paper is to draw attention to a second, unofficial Hobbesian theory, which is strikingly similar to that of the *later* Wittgenstein. However, my task has been complicated by Isabel Hungerland's recent reinterpretation of Hobbes's official theory itself as anticipating, not just the later Wittgenstein, but the linguistics of H. P. Grice. In her essay, 'Hobbes's Theory of Language, Speech and Reasoning',[1] she challenges the traditional interpretation on the following four points in particular.

1. On the traditional interpretation, Hobbes holds that names are names of mental images; for Hungerland, names denote things, not mental images.

2. On the traditional interpretation, names are arbitrarily chosen without there being any reason in nature for our classifications; for Hungerland, our classifications have reasons, namely 'similitudes' between things; and significant speech 'conveys' these reasons to the hearer.

3. On the traditional interpretation, the primary function of language is to register thoughts, and its function of signifying thoughts to others is only secondary; for Hungerland, signification consists in speech-acts presupposing a community of language-users, and this function of language is logically prior to that of registering thoughts.

4. On the traditional interpretation, language is the differentia between humans and animals; for Hungerland, 'Hobbes's theory of communication . . . aims to cover the whole range . . ., from the howls and cries of wild creatures through the learned performance of chimps to the discourse of geometers' (p. 134).[2]

The divergence between these interpretations of one and the same text is remarkably wide; and Hungerland is well aware that this divergence stands in need of some explanation. In the section immediately following, I shall discuss the plausibility, or otherwise, of Hungerland's general explanation of the divergence. In the section after that, I shall consider her specific arguments in detail, and in the course of doing so I shall expound the main features of Hobbes's official theory itself. In the last two sections, I shall argue that Hobbes did have a theory similar to the one Hungerland attributes to him, but that it is to be found in *Leviathan*, rather than in *De corpore*; and finally I shall tackle the question of how this theory is to be reconciled with Hobbes's official theory.

3.2 *The 1656 translation of De corpore*

Hungerland's principal explanation for the divergence between her own and the traditional interpretation of Hobbes is that he was *mistranslated*. Until her commissioning of Martinich's new translation, the only one available was the anonymous translation of 1656,[3] which appeared a year after the first Latin edition. Her position is that the 1656 translation was so bad that it turned Hobbes's meaning completely on its head.

Let us suppose for a moment that the translation was indeed a very bad one. This might explain why Hobbes has been misunderstood by non-Latinate English-speaking philosophers, of whom there are now a large number. There are, however, at least three things which it cannot explain.

First, why did Hobbes fail to take the opportunity to correct such a travesty of his theory? In the Epistle to the Reader, his translator writes:

> I thought fit before I published [my translation] to pray Mr Hobbes to view, correct, and order it according to his own mind and pleasure. Wherefore, though you find some places enlarged, others altered, and two chapters, XVIII and XX almost wholly changed, you may nevertheless remain assured, that as now I present it to you, it doth not at all vary from the author's own sense and meaning.[4]

Hungerland sees this difficulty, and says that the third of three

> circumstances . . . responsible for the widespread misinterpretation of Hobbes's philosophy of language by eminent scholars . . . is Hobbes's general carelessness, shown, for example, in his approving of a translation full of mistakes. (p. 23)

If the translator is telling the truth about the extent to which Hobbes himself altered chapters XVIII and XX and other passages, then

Hobbes's 'carelessness' must have been remarkably selective.

Let us, however, suppose that Hobbes was indeed selectively careless. This still leaves the second difficulty, which is that of how the translator himself, together with all other readers of Hobbes's Latin, managed to misunderstand him in the first place. For example, Hungerland praises Leibniz, 'who most clearly recognized the originality of Hobbes's entire system of ideas' (p. 23); but even Leibniz failed to detect any of the Gricean aspects of Hobbes's theory. Hungerland has anticipated this objection too. Another reason she gives for the widespread misinterpretation of Hobbes is:

> Hobbes's originality itself. All of us, including historians of philosophy, have a tendency to overlook patterns with which we are not familiar. Philosophy of language as a recognized discipline is a recent one. Hobbes's views in this area, so far ahead of his time, were understandably overlooked. (pp. 22–3).

Elsewhere she explains that Hobbes has traditionally been interpreted in the light of the linguistic theories of the logical positivists and the early Wittgenstein (p. 153, note 3), but that we no longer have any excuse for interpreting him this way:

> By those who have grossly misinterpreted him, Hobbes has been taken as holding a theory of language and meaning which – at least in the crude form attributed to him – has been discredited for more than a decade. (p. 15)

Now, this may well explain why Hobbes has been seen as a fellow spirit by philosophers and linguists operating in a predominantly positivist and materialist climate, such as prevailed for a while in the Anglo-Saxon world; but it cannot account for *earlier* misunderstandings of Hobbes. Philosophers like Leibniz and Berkeley were *unusual* in agreeing with their version of Hobbes. The vast majority of his contemporaries and successors, so far from reading familiar patterns into his Latin text, found it absurd and heretical. Indeed, many of Hobbes's detractors held views closer to Hungerland's Hobbes than their own – particularly in relation to universals, and the conveying of reasons by significant speech.

The third difficulty is that Hungerland finds her own version of Hobbes's theory only in his allegedly mistranslated Latin. She sees no trace of it in *Leviathan*,[5] which Hobbes was working on at the same time, and of which ch.iv is a summary of *De corpore* I.ii. Hungerland does not even face the problem of why Hobbes should have concealed his true meaning from English-language readers – a problem which is especially acute, given that a Gricean theory would be much less dangerously positivist and materialist than the theory sketched in *Leviathan*.

Nevertheless, it is just conceivable that, on this one issue, Hobbes chose to express his genuine theory only in Latin, and wrapped it up so well that it has played Sleeping Beauty to Hungerland's Prince Charming. But if we are to be persuaded of this, we need some positive evidence that *De corpore* was mistranslated, otherwise the obvious conclusion will be that Hungerland rather than everyone else has misconstrued Hobbes's Latin. In general, there would seem to be three possible ways of showing that a translation is a *mis*translation.

The most natural approach would be to appeal to scholarly evidence of how words were used at the time. However, this move is not open to Hungerland, since she makes the opposite claim, that contemporary usage itself blinded people to Hobbes's own, original distinctions. Thus Hungerland admits that, at the time, both *denotare* and *significare* meant simply 'to mean', without implying her contrast between denotation and signification (p. 153, note 2); and Martinich notes that *oratio* normally meant 'sentence' rather than his 'speech-act' (p. 350).

A second approach would be to claim that Hobbes's explicit definitions were ignored. Now, Hobbes did indeed define some of his basic terms, such as *nomen, nota*, and *signum;* but no-one is accusing the translator of getting *these* wrong. Unfortunately for Hungerland, the terms she takes as crucial were not explicitly defined by Hobbes; and she herself scolds him for his carelessness (again!) in establishing the distinction between denotation and signification only on one occasion, and then only by implication:

> He establishes, by implication, the difference between 'to denote' (*denotare*) and 'to signify' (*significare*). Hobbes, having given his interpretation of the locution in question, continues, somewhat carelessly, to employ it in subsequent passages without reminding the reader how it is to be taken. (p. 34)

The only remaining approach would be to show that a given translation makes worse sense in the context than one's own proposed alternative – in other words, to appeal, not to what a writer *said*, or to what he *said he meant*, but to what he *must have* meant. This sort of move can be perfectly legitimate; but it has to be remembered that it reverses the logical relationship between text and theory: it is no longer the case that the text uniquely determines the theory, but that the theory determines the meaning of the text. In order to avoid begging the question of what the theory was, there must be independent criteria, such as internal consistency, obvious truth, or contemporary relevance. In her final paragraph, Hungerland comes close to recognising that this is the status of her claim that Hobbes has been mistranslated, and hence that it can only be a corollary of her interpretation, rather than a premise from which it follows. She writes:

> In conclusion, we wish to point out that our reading of Hobbes's theory of speech and language is not only consistent with the corrected text [i.e. corrected in the light of their reading], but that it shows his theory to be consistent in the following way. First, . . . the theory seems to be self-consistent. Second, it shows his theory to be . . . an integral part of his system. . . . Third, it shows that the same methodological principles . . . are at work in this area of his system as in other areas. Finally, our reading makes it plain that Hobbes's thought . . . is . . . original, bold, and highly relevant in certain respects to contemporary [i.e. modern] theory. (p. 148)[6]

So, since issues about the correct translation of Hobbes's text cannot be settled independently of the question of what philosophical theory he held, I shall proceed directly to an examination of the four main points of divergence between Hungerland's interpretation and the traditional one.

3.3 *Hungerland's misinterpretations of Hobbes's official theory*

3.3.1 *Names as denoting things*

According to the Hungerland version of Hobbes, names denote things, whereas signs (or vocal sounds) signify concepts. This implies a mutual mapping of three pairs of contrasted terms: name/sign, denotation/signification, and thing/concept.

The name/sign contrast is pure fiction, since signs are a sub-class of names, and contrasted with *notes* (I.ii.3–5).[7]

Hungerland herself admits that 'denote' is not a Hobbesian technical term, since in I.vi.11 he talks of words 'denoting' concepts (p. 153, note 2; p. 154, note 18). But 'signify' is in the same position, since Hobbes talks of names signifying (e.g. the last sentence of I.ii.3), and he even implies that notes signify: in *Leviathan*, ch. iv, the first abuse of speech, which applies specifically to the function of words as notes, involves 'inconstancy of the signification of [men's] words'. It is true that Hobbes sometimes distinguishes *notare* and *significare*, meaning 'to be the note of' and 'to be the sign of' (e.g. I.ii.3); but in general *denotare* and *significare* mean 'mean' indifferently.

As for the grafting of the spurious denotation/signification distinction onto that between things and concepts, the source of this seems to lie in the section headings of I.ii.5 and 6, namely: 'Names are signs not of things but of thoughts', and 'Names, what things they are of'.[8] At first sight there seems to be a straight contradiction between names *not* being of things, and names *being* of things. Hungerland's solution is to interpret the first heading as saying that signs are of thoughts, and the second as saying that names are of things. If this is correct, then Hobbes is unbelievably careless with the word 'name', using it first as *in*cluding signs, and then as *ex*cluding signs. Hungerland then quotes the words of the 1656 translation:

> For . . . these, *a man*, *a tree*, *a stone*, are the names of the things themselves.[9]

She concludes that names *name* things themselves, but *are signs of* the concepts of things. It is difficult for the post-Kantian reader not to see a contrast here between things themselves, and concepts of things. But the quotation is torn from its context, and *ipsae res* means little more than '*things*' italicised.

In order to sort out the apparent contradiction between the two section headings, I shall offer my own translations. Section 5, that 'Names are signs not of things but of thoughts', runs as follows:

> But since, as I have stated, names marshalled into sentences are signs of conceptions, it is obvious that they are not signs of *things*. The only sense that can be made of saying that the sound of the word 'stone' is a sign of a *stone*, is that someone hearing the word would gather that the speaker had been thinking about a stone. Consequently, the old argument as to whether names signify *matter*, *form*, or a *compound*, and other such metaphysicians' arguments, are for people who are completely off the track, and do not even understand the words they are arguing about.[10]

In section 3, Hobbes has already said that individual names are not signs, and that we have to use whole sentences to communicate our thoughts to others.[11] Here he grudgingly allows *some* sense to the idea of an individual word functioning as a sign: not the sign of a *thing*, as in the examples of section 2, where a cloud is a sign of rain, or ivy is a sign of wine; but only as a sign that there has been a certain thought. Since names do not signify *things, a fortiori* they do not signify metaphysical *components* of things. So, since he has already said that notes are only of thoughts, and that sentences are signs only of thoughts, we now have an unambiguous statement, *pace* Hungerland, that names in *all* their functions are of thoughts, not of things.

As for section 6, 'Names, what things they are of', the trouble is caused, not by the word 'name', but by the word 'thing'. In Latin, as also sometimes in English, the word *res* or 'thing' has an indispensible grammatical function. As it happens, we can say in English: 'Names, what they are of;' but in Latin, plain *quorum* would make no sense, and Hobbes has to say *quarum rerum*. But far from saying that names are of *things*, the purpose of the section is to point out that *non-things* also have names:

Nor is it necessary for every name to be the name of some actual thing. For in the same way as 'man', 'tree', and 'stone' are the names of *things*, so dream-images of a man, a tree, or a stone each have their own names, even though they are not things, but only fictions and illusions of things. It is a fact that they are remembered, so they must be noted and signified by names, no less than actual things are.[12]

The contrast here is between waking and dreaming, and there is not the slightest suggestion that when we are awake we talk about things themselves, rather than about conceptions or sensory images.

So, what Hobbes is saying in sections 5 and 6 is that all names refer to thoughts, whether functioning as notes or as signs; and that these thoughts are not necessarily of actual things.

3.3.2 *Signs as conveying reasons based on similitudes*

In Hungerland's own words:

An important consequence of Hobbes's account of n-expressions [names] and of acts of signifying is that the reason for applying a given n-expression is not denoted or stated in any communication in which the n-expression is employed. The reason, one might say, is *conveyed* in the act of speech and conveyed because to be a member of a linguistic community requires, among other things, an understanding of the reasons for the community practice in applying n-expressions. (p. 57)

Elsewhere, Hungerland tells us that it is intensions as opposed to extensions that are signified,[13] and that intensions are concepts of real 'resemblances', 'similarities', or 'similitudes', which ground our use of universal terms.[14] She thus rejects the traditional view that Hobbes was an extreme nominalist (p. 18), and sees him rather as an Aristotelian (p. 102). I shall not here enter into any discussion of Hungerland's anachronistic use of the intension/extension contrast, nor of her attribution to Hobbes of a belief in the power of words to 'convey reasons'. Both depend on her interpreting Hobbes's account of universals as realist.

Hungerland's sole source for saying that Hobbes grounded universals in 'similitudes' is a passage from *De corpore* (I.vi.11), where he says that someone without language would not know whether or not a property he perceived in one triangle was in another triangle *dissimilar* to it, or even in the same one in a different position (p. 81). However, Hobbes does not merely fail to use the word 'similitude' as such, but he is using the word 'dissimilar' in the geometricians' technical sense, according to which two figures are 'dissimilar' if they fail to have certain *identical* properties in common. Hobbes is not talking about dissimilarity in the ordinary, vague sense.

For a use of the word 'similitude' itself, Hungerland could have appealed to I.ii.7, where Hobbes writes:

> The word 'man' denotes any of many men, and the word 'philosopher' denotes any of many philosophers, on account of the similitude of all of them.[15]

But here too, far from producing a 'similitude' theory of universals, he is contrasting 'positive' names, which are based on complete or partial identity ('similitude, equality, or identity'),[16] with 'negative' names, such as 'not-white'. Similitude, or sharing an identical property, is only one of a number of bases for classification. Again, in *Leviathan*, ch. iv, p. 13,[17] Hobbes writes:

> One Universall name is imposed on many things, for their similitude in some quality, or other accident: And whereas a Proper Name bringeth to mind one thing onely; Universals recall any one of those many.

But even here, Hobbes does not base his theory on some notion of 'similitude', but on that of an identical quality or accident. Like so many other philosophers of his day, he thought of qualitics as nameable particulars which recurred in experience, so that there was no need for any realist theory of universals to account for their repeated recognition.[18] One might well sympathise with

Hungerland's refusal to accept Hobbes's talk of individuals sharing the same quality as an adequate account of universals; but this has nothing to do with the historical question of whether or not he was conscious of the same problems as Hungerland.

A final consideration is the following, delightful passage from *Leviathan*, ch. viii, pp. 39–40.

> And particularly, in the question of Transubstantiation; where after certain words spoken, they that say, the White*nesse*, Round*nesse*, Magni*tude*, Quali*ty*, Corruptibili*ty*, all of which are incorporeall, *&c.* go out of the Wafer, into the Body of our blessed Saviour, do they not make those *Nesses*, *Tudes* and *Ties*, to be so many spirits possessing his body? For by Spirits, they mean alwayes things, that being incorporeall, are neverthelesse moveable from one place to another. So that this kind of Absurdity, may rightly be numbred amongst the many sorts of Madnesse.

In the light of this, is it really conceivable that Hobbes could have grounded his account of universals in a 'Tude', such as 'similitude'?

3.3.3 *Signification as prior to naming*

The focal point of Hungerland's interpretation of Hobbes is her reading him as making signification in a community prior to private acts of naming. She sums this up in the slogan: 'no meaning without communication' (p. 37). She frequently *asserts* her position, but actual arguments are curiously thin on the ground.[19] The nearest I can find to an argument is an assumption that when Hobbes uses terms such as *vox* or *oratio*, as opposed to, say, *verbum* or *propositio*, it must be significant that the former pair are etymologically related to the notion of public speaking, and hence presuppose a community of language-users.

My reply to this is that the etymological significance of such terms had long since been lost, so that by Hobbes's time they meant no more than 'word' and 'sentence' respectively. If he had wanted to coin the concepts of a 'vocal sound' or a 'speech-act', as Hungerland would have us believe, he would have had to do so explicitly. As it happens, he does sometimes distinguish a vocal sound from a word, and a speech-act from a sentence, but not by using Hungerland's spurious contrasts. Thus in I.ii.5 he refers to 'the sound of the word "stone" ', using the expression *sonus vocis*. The Hungerland/Martinich translation, 'the sound of the vocal sound "stone" ' (p. 201), merely converts sense into nonsense. Again, in I.ii.4, Hobbes talks of a 'word [*vox*] . . . marshalled into a sentence [oratio], and uttered to others', carefully distinguishing the processes of sentence construction and of vocal expression. Martinich renders the passage as: 'a vocal sound . . . ordered in speech and uttered to others' (p. 199); but his mistranslation of *vox* as 'vocal sound', and of *oratio* as 'speech', compounded by his dropping of a crucial comma, destroys the contrast, and makes the qualification 'and uttered to others' completely redundant.

A similar etymological assumption is made on pp. 55–6, where, after admitting that Hobbes sometimes uses the word *denotare* 'for the relation between things and things named, considered apart from the signifying situation', (i.e. in contravention of Hungerland's insistence that there is 'no meaning without communication'), she then continues:

In other passages, however, Hobbes employs terminology which points to the dependence of n-expressions [names], as parts of the language, on a community, on signifying, or on speech situations. Thus, sometimes in discussing names and things named, Hobbes employs the passive of 'to call', (*vocare* or *appellare* in the Latin), speaking of A's being called 'A'. Things, of course, are called whatsoever they are called by people. In Hobbes's theory, any such persons must be members of a linguistic community (at least two members required) and objects are called 'so-and-so' as a result of a decision to establish the correlation in question between certain vocal sounds (types) and certain items in the world.

It is of course true that there have to be people for things to be 'called' anything; but it is no less the case that there has to be at least one person to establish a relation between a name and a thing named, so that names can denote. Since there is nothing special about the words *vocare* or *appellare* to prevent them from applying to a private baptism by a solitary language-user, Hungerland's contrast with *denotare* falls to the gound. Language presupposes a language-user; but not, in Hobbes's view, a community of language-users.[20]

This brings me to a more serious obstacle to Hungerland's interpretation than mere lack of evidence. In *De corpore* I.ii.3, Hobbes makes it as clear as language could, that noting is logically prior to signification. He writes:

But since, as we have said, philosophy requires both *notes and signs* (*notes* so that we can remember our thoughts, and *signs* so that we can make them known), names do both these things. But their function as *notes* is prior to their function as *signs*. For even if a man existed alone in the world, he would find them useful as an aid to his memory, even though they could be of no use for making his thoughts known without anyone to make them known to.[21]

How does Hungerland get round this? On pp. 25–30 she explains how the Paduan methodology, which Hobbes had adopted, involved simplified fictions intended to bring out logical, rather than historical, priority. She then concludes, on p. 33, that the sequence indicated by the *before* and *after* in I.ii.3 'is not temporal in nature, but logico-methodological'. But this does not yet solve the problem, since no-one would interpret Hobbes's assertion as being about historical, rather than logical, priority. Hungerland comes back to the question on pp. 42–6, and there gives a double answer. She writes as follows:

The significance of the *before* and *after* in question is double. First, there is the methodological *before* and *after*, involved in the priority of the analytical stage of the Paduan method. Second, if we consider the relation between the notion of discovering something, *that p*, and that of teaching it, it is clear that there is a logical priority of the first – that is to say that if one is *to teach that p*, one must first *find out that p*. (p. 46)

So Hungerland's first explanation is that Hobbes is saying that he is beginning his *analysis* with notes, in order to discover later that signs are logically prior to notes. But this interpretation is completely absurd, since Hobbes says explicitly that it is the *actual function* of notes themselves which is prior. Furthermore, Hobbes reinforces this interpretation, in the penultimate sentence of the section, where he expresses the same position, only now in terms of what is primary or essential, as contrasted with what is accidental or non-essential:

Therefore the nature of a *name* consists primarily in the fact that it is a *note* used as an aid to memory; but it also serves the accidental function of signifying and making known the things we retain in our memories.[22]

The second part of Hungerland's explanation is that Hobbes is saying that we have to learn facts before we teach them to others. But where does this reference to *teaching* come from? It may be that she has been misled by the English translation, which renders the Latin *demonstrare*, here meaning to 'make known', as to 'teach' (quite correctly in seventeenth-century English). But Hobbes insists, in this very same paragraph, that notes are of individual thoughts, not of complete propositions. So when he says we use notes for recording our thoughts, he cannot possibly mean that we are 'finding out that p' before teaching someone else that p, since 'p' would be a complete proposition.

On the other hand, Hungerland may have in mind a different private-language passage, namely *De corpore* I.vi.11. Here Hobbes talks of a solitary person, who is aided by notes in the discovery of geometrical proofs, but who cannot teach what he has discovered without using words as signs. Elsewhere, Hungerland explains this as another 'Paduan fiction' (pp. 25, 42–5, 80–83): the private language which Hobbes describes is not a logically more basic form of language, but a 'curiously limited use' (p. 25) of the language we already have. That is what Hungerland says. But Hobbes ends his section as follows:

> Therefore the solitary person can become a philosopher without a teacher. Adam had this capacity. But to teach, that is to produce proofs, requires two people, and logically connected discourse.[23]

I do not wish to get embroiled in the question of whether or not Hobbes was sincere in his appeals to scriptural authority. Martinich, for example, refers to Hobbes's discussion of what happened at Babel as 'comic relief' (p. 351). Either way, Hobbes's message is clearly that a solitary philosopher with his own private language is a genuine possibility, not a mere 'Paduan fiction'.

3.3.4 *Language as in common with animals*

Finally, it is one of Hobbes's fundamental theses that the differentia between human beings and animals is not consciousness, or the soul, or reason, but *language* (cf. *Leviathan*, ch.ii, p. 8). How does Hungerland reconcile this with her claim that, for Hobbes, animals have language? She even refers (pp. 47–50) to *De homine*, x. 1, of which the heading is: 'Language defined; that it is peculiar to humans'.[24] The source of this contradiction is that Hungerland treats *signs* rather than *notes* as essential to language. Hobbes does indeed allow that animals have the capacity to signify; but it should by now be clear that for Hobbes it is in fact *notes* that are essential to language, not *signs*.

Hungerland tries to take into account the difference that only humans invent arbitrary signs (p. 49); and more than once she refers to this as 'the great break between men and animals' (pp. 32, 49). But what precisely is meant by a 'great break'? Clearly, it must be big enough to accommodate Hobbes's difference in kind, but without being so great as to falsify Hungerland's essential continuity. What Hobbes actually says is that the difference is precisely the difference between having and not having a language. Although both humans and animals can signify, only humans have language, or *sermo*, because animals cannot use 'words as words, but only as signs'.[25] What animals lack is the capacity to use words as *notes* – the creative activity of giving names to one's thoughts.

Children are in the same position as animals, since their naming capacity is still only dormant, and they acquire language from adults. As Hobbes says in *Leviathan*, ch. v, p. 21:

> Children therefore are not endued with Reason at all, till they have attained the use of Speech: but are called Reasonable Creatures, for the possibility apparent of having the use of Reason in time to come.

While children are learning to speak, their communication depends on learned associations between words and thoughts, with the words functioning as if they were natural signs.[26] Human reason arises only with the capacity to give arbitrary names to one's private thoughts.

3.4 *The ordinary-language theory in Leviathan*

I trust that I have succeeded in showing that Hobbes's official theory of meaning is as positivist as it has always been taken to be. So far from anticipating the later Wittgenstein or Grice, Hobbes explicitly espouses the thesis that ordinary language is parasitic upon the capacity to invent a private language. I shall now show that Hobbes had a second, unofficial theory, which happens to be quite close to the theory attributed to him by Hungerland. It does not occur in *De corpore*, where Hungerland thought she had found it, but only in *Leviathan*, which she almost completely ignored.

The essence of this theory is that an expression is meaningful if it belongs to a vernacular language, such as ordinary people use in real life, as contrasted with the 'barbarous' Latin of philosophy professors and others.[27] Hobbes's general position is clear enough to need no re-interpretation in the light of some theory about what he must really mean, so I shall simply reproduce a couple of passages in which he uses this remarkably Wittgensteinian criterion of meaning as a weapon against scholasticism. First, from chapter viii, p. 39:

> There is yet another fault in the Discourses of some men; which may also be

numbred amongst the sorts of Madnesse; namely, that abuse of words, wherof I have spoken before in the fifth chapter, by the Name of Absurdity. And that is, when men speak such words, as put together, have in them no signification at all; but are fallen upon by some, through misunderstanding of the words they have received, and repeat by rote; by others, from intention to deceive by obscurity. And this is incident to none but those, that converse in questions of matters incomprehensible, as the Schoole-men; or in questions of abstruse Philosophy. The common sort of men seldome speak Insignificantly, and are therefore, by those other Egregious persons counted Idiots. But to be assured their words are without any thing correspondent to them in the mind, there would need some Examples; which if any man require, let him take a Schoole-man into his hands, and see if he can translate any one chapter concerning any difficult point; as the Trinity; the Deity; the nature of Christ; Transubstantiation; Free-Will, *&c.* into any of the moderne tongues, so as to make the same intelligible; or into any tolerable Latine, such as they were acquainted withall, that lived when the Latine tongue was Vulgar. What is the meaning of these words. *The first cause does not necessarily inflow any thing into the second, by force of the Essentiall subordination of the second causes, by which it may help it to worke*? They are the Translation of the Title of the sixth chapter of *Suarez* first Booke, *Of the Concourse, Motion, and Help of God.* When men write whole volumes of such stuffe, are they not Mad, or intend to make others so?

The second passage is from chapter xlvi, p. 379:

More examples of Vain Philosophy, brought into Religion by the Doctors of Schoole-Divinity, might be produced; but other men may if they please observe them of themselves. I shall onely adde this, that the Writings of Schoole-Divines, are nothing else for the most part, but insignificant Traines of strange and barbarous words, or words otherwise used, then in the common use of the Latine tongue; such as would pose Cicero, and Varro, and all the Grammarians of ancient Rome. Which if any man would see proved, let him (as I have said once before) see whether he can translate any Schoole-Divine into any of the Modern tongues, as French, English, or any other copious language: for that which cannot in most of these be made Intelligible, is not Intelligible in the Latine. Which Insignificancy of language, though I cannot note it for false Philosophy; yet it hath a quality, not onely to hide the Truth, but also to make men think they have it, and desist from further search.

3.5 *The relation between the two theories*

The modern reader may find it odd that Hobbes should have adhered to two separate theories of meaning simultaneously. We are used to philosophers providing unitary theories, which perform the double function, both of explaining what it is for expressions to have meaning, and of providing a criterion for demarcating meaningful from meaningless forms of discourse.[28] For example, Locke, Berkeley and Hume all analysed meaning in terms of a relationship between words, and units of experience; and they outlawed more and more of dogmatic metaphysics, on the grounds that ideas corresponding to metaphysicians' words could never have entered their minds. More recently, the logical positivists used the verification principle to eliminate metaphysical, religious and moral discourse; and Popper used his falsifiability criterion to demarcate genuine from pseudo-science. At the other end of the spectrum, there is the school influenced by the later Wittgenstein, which has analysed meaning in terms of the use of linguistic expressions within a community. This more sociological approach has tended to be tolerant towards everyday discourse on any topic (including religion and morality),[29] but to make abstract and theoretical uses of language in their turn problematic, or even actually meaningless.

Hobbes's dual theory therefore looks highly appealing: instead of indulging in wholesale rejection of large areas of *prima facie* meaningful discourse, he provides a complementary pair of criteria. The official criterion in *De corpore* gives conditions for the meaningless of abstract, theoretical and technical language, namely that arbitrary definitions should be adhered to consistently; and the unofficial criterion in *Leviathan* requires adherence to the established usage of ordinary language, in spheres such as everyday life, politics, morality and literature.[30]

More recent writers have taken sides in the modern conflict between the 'two cultures'; but Hobbes was living in an age when any line of demarcation was drawn very differently. In the seventeenth century, the main divide was not between science and the humanities, but between the modernists and the scholastics. The modernists certainly included those we would now classify as scientists and technologists. But they also included (often in the same persons) the students of Classics and modern literature, who held that the best way to make progress in all areas of knowledge was to stand on the shoulders of the ancients; to think and write in a Latin purified of medieval barbarisms, or in the vernacular; and to adopt the canons of reasoning developed by the ancient rhetoricians, rather than the syllogisms and disputations of the scholastics.[31]

At first, the modern movement existed almost entirely outside the university system, so that the cultural divide was effectively between university professionals, and non-university amateurs, such as Hobbes himself. As universities gradually adopted the new ideas, the Arts faculties generally moved first, and sponsored both humanism and the new science in opposition to the traditionalism of the higher faculties of Theology, Law, and Medicine. Hobbes was therefore reflecting contemporary institutional realities, when he championed *both* positive science *and* the humanities (especially Classics, rhetoric, and the use of the vernacular) against the entrenched power of traditional scholasticism.[32]

Nevertheless, despite the obvious advantages of a dual theory of meaning, especially in Hobbes's particular historical context, there is a considerable difficulty over establishing a plausible line of demarcation between their spheres of influence, so as to avoid ambiguity or conflict. Since Hobbes gave only one of his criteria official status, it is hardly surprising that he did not directly confront the question of the relation between the two; nor is it surprising to find hints that he would have solved the problem by making the unofficial criterion subordinate to the official criterion.

One such hint is that, by contrast with his orderly and serious exposition of the official theory of meaning, Hobbes appeals to the unofficial criterion only incidentally, and mainly in order to poke fun at the scholastics. For example, the following passage from the end of *Leviathan*, ch. i, p. 4. is clearly intended as a piece of rhetoric, rather than as a serious exposition of scholastic doctrine as such.

But the Philosophy-schooles, through all the Universities of Christendome, grounded upon certain Texts of *Aristotle*, teach another doctrine; and say, For the cause of *Vision*, that the thing seen, sendeth forth on every side a *visible species* (in English) a *visible shew*, *apparition*, or *aspect*, or *a being seen*; the receiving whereof into the Eye, is *Seeing*. And for the cause of *Hearing*, that the thing heard, sendeth forth an *Audible species*, that is, an *Audible aspect*, or *Audible being seen*; which entring at the Eare, maketh *Hearing*. Nay for the cause of *Understanding* also, they say the thing Understood sendeth forth [an] *intelligible species*, that is, an *intelligible being seen*; which coming into the Understanding, makes us Understand. I say not this, as disapproving the use of Universities: but because I am to speak hereafter of their office in a Commonwealth, I must let you see on all occasions by the way, what things would be amended in them; amongst which the frequency of insignificant Speech is one.

However, the fact that the style of this passage is highly satirical in no way detracts from the validity of Hobbes's point, that the scholastic doctrine of intensional species makes no sense, *either* as defined technically, *or* allowing the words their vernacular meaning. The latter alternative is in no way subordinated to the former.

A second reason for supposing that the unofficial theory might be subordinate comes in the passage from ch.viii, quoted in Section 3.4 above, where Hobbes says that the common sort of men *seldom* speak insignificantly. Similarly, in the second quotation, from ch.xlvi, he specifies that the vernacular into which scholastic Latin is to be translated must be *copious*.[33] Both these qualifications imply some separate criterion for deciding those cases when ordinary people do speak insignificantly, and for deciding what is to count as a sufficiently 'copious' language. However, it is not the case that this separate criterion is the *official* theory of meaning, which would then reduce the unofficial theory to a mere rule of thumb. Rather it is a question of the proper *scope* of each theory. The ordinary-language criterion is always supreme in its own sphere; but the language that happens to be spoken by ordinary people can lapse into meaninglessness if contaminated by ill-defined or misunderstood terms belonging properly to the technical sphere.[34] Thus in *Leviathan*, ch.iv, p. 12, Hobbes distinguishes ordinary language, useful technical language, and meaningless technical language; and implies by his silence that there could be no such thing as meaningless ordinary language.

And so by succession of time, so much language might be gotten, as [Adam] had found use for; though not so copious, as an Orator or Philosopher has need of. For I do not find any thing in the Scripture, out of which, directly or by consequence can be gathered, that *Adam* was taught the names of all Figures, Numbers, Measures, Colours, Sounds, Fancies, Relations; much less the names of Words and Speech, as *Generall*, *Speciall*, *Affirmative*, *Negative*, *Interrogative*, *Optative*, *Infinitive*, all of which are usefull; and least of all, of *Entity*, *Intentionality*, *Quiddity*, and other insignificant words of the School.

But if ordinary language is autonomous within its proper sphere, how can it become contaminated? Here Hobbes is driven to a conspiracy theory: left to nature, everything would be all right; but academics, priests and politicians corrupt natural language by importing meaningless jargon, and use this to preserve their power over ordinary people. For example, they frighten ordinary folk with stories about immaterial ghosts and demons, and then offer protection, at a price, against the evils they are imagined to cause. As he writes in *Leviathan*, ch. iv, p. 15:[35]

In wrong, or no Definitions, lyes the first abuse [of speech]; from which proceed all false and senseless Tenets; which make those men that take their instruction from the authority of books, and not from their own meditation, to be as much below the condition of ignorant men, as men endued with true Science are above it. For between true Science, and erroneous Doctrines, Ignorance is in the middle. Naturall sense and imagination, are not subject to absurdity. Nature it selfe cannot erre: and as men abound in copiousnesse of language; so they become more wise, or more mad than ordinary. . . . For words are wise mens counters, they do but reckon by them: but they are the mony of fooles, that value them by the authority of . . . but a man.

So natural language is in itself infallible – 'Nature cannot err': but ordinary people can speak insignificantly, as soon as they start talking about theoretical matters beyond the proper scope of ordinary language.

A more serious difficulty involved in regarding Hobbes's two criteria as of equal status, is that, in the quotation from ch.viii, he says that failure to pass the ordinary-language criterion shows that 'words are without any thing correspondent to them in the mind'. It seems to follow from this that he really has only one *definition* of meaning: in both theoretical speech and ordinary language, for a word to be meaningful is for it to correspond to something in the mind. So how can there be two *theories*? Nowadays we tend to deny the possibility of any logical gap between a definition of a term, and criteria for its use; but for Hobbes, the connection is clearly contingent. In his official theory, he simply assumes without question that whenever a theoretical term is used consistently, there will be a corresponding thought; and in his unofficial theory he dogmatically asserts that nature cannot err – that as long as ordinary language is used in accordance with nature, the right thoughts will always be present to the mind. The connection is therefore weaker than it is for, say, Norman Malcolm, who takes it as logically sufficient for meaning, that 'this language game is played!'[36] For Hobbes, the connection is guaranteed by Nature rather than by logic.

Given Hobbes's gap between definition and criteria, it is up to us to decide whether there is one theory or two. The obvious choice is to say that there are *two* theories, since the two criteria can function without the definition. The definition, on the other hand, has no work to do, since we can never make any direct check on the contents of other people's minds when they use words, meaningfully or otherwise.

A final consideration in favour of interpreting Hobbes as having two theories of meaning arises from the relation between language and politics. One distinction between the scope of the two criteria is that the official theory applies primarily to the *private* domain, and the unofficial one to the *public* domain. The official theory starts out from the scientist or philosopher thinking by himself, and then promulgating his technical vocabulary so as to become accepted by the intellectual community. But in the case of natural language, all the important word-creation and agreeing on words was done long ago. Consequently, ordinary people are not normally called upon to exercise their capacity to notate thoughts. In order to communicate successfully, they have to accept a ready-made vocabulary; and so in practice their language differs only in degree rather than in kind from animal communication, which consists exclusively of signs.[37] This is one reason why ordinary folk are so ready to take on trust the linguistic authority of intellectuals, and why they are so vulnerable to being misled.

A principal purpose of *Leviathan* was to prevent religious war, by promoting the sovereign as the ultimate linguistic authority in the public domain. In particular, the meaning of Holy Scripture would be uniquely fixed by the sovereign operating in accordance with the principles of natural usage, and outlawing the contrary interpretations of academics, priests, and religious enthusiasts. As it happens, Hobbes does not say whether the sovereign has power over *all* public language, or only in the political and religious spheres. Thus in the transition between chapters xxxiii and xxxiv of *Leviathan* (pp. 206–7) he writes:

> . . . wee are to judge of the Authority of Interpreting the Scripture. For, whosoever hath a lawfull power over any Writing, to make it Law, hath the power also to approve, or disapprove the interpretation of the same.
>
> Seeing the foundation of all true Ratiocination, is the constant Signification of words; which in the Doctrine following, dependeth not (as in naturall science) on the Will of the Writer, nor (as in common conversation) on vulgar use, but on the sense they carry in the Scripture . . .

However, supposing a sovereign who had the powers Hobbes would like him to have, it is difficult to see why he should not be able to interfere with ordinary language by adding what he wanted to it, as the academics and priests had already done in their own interests. The additions might be meaningful, or they might not; and his motives might be to promote the common good, or they might not. Either way, his power to influence language would be far greater than that of any sectional group. There is also nothing in Hobbes to prevent the sovereign from going in the other direction of making language *less* 'commodious', and hence limiting the expressibility of dissident thoughts, as with the Newspeak of Orwell's *1984*.

But just as Hobbes allows individuals freedom of conscience in religion and politics, provided only that they consent to the sovereign in public; similarly, his official theory of meaning would guarantee at least the possibility of freedom of thought under a state of linguistic totalitarianism. This is an escape which would not be available in a Wittgensteinian world in which a central authority wrote the rules of all the language-games. Hobbes's double requirement, that control of public speech is the responsibility of the state, and that the individual has complete freedom of thought, can be satisfied only if there are two autonomous ways in which words can have meaning – that is, only if he has two, parallel theories of meaning.

Notes

1 This essay, written in collaboration with George Vick, forms the introduction to Aloysius Martinich's new translation of the first part of Hobbes's *De corpore*, New York, 1981, to which Martinich appends his own commentary, defending the same interpretation. Since Hungerland is the protagonist of the trio, I shall refer to the interpretation simply as hers.

2 Page numbers refer to the Hungerland–Vick–Martinich volume. I make no comment on Hobbes's alleged aim to account for the linguistic feats of chimpanzees.

3 Reprinted in Sir William Molesworth, ed., *English Works*, London, 1839.

4 *Ibid*, note 3, p. vi.

5 It is remarkable that Hungerland mentions *Leviathan* only once, and then only in connection with Hobbes's uncontroversial observation that not all sentences assert propositions, see p. 64, cf. pp. 149–52.

6 Similar criteria are hinted at on p. 23: 'Where we have had to fill in gaps in the statement of his theory, we have done it on the basis of the assumption that Hobbes employs the same method of theory construction throughout his work.' Cf. p. 129: 'We have not hesitated to fill in gaps in [Hobbes's] theory [of speech and language], employing certain accepted

principles of interpretation.' They do not, however, state what these principles are.

7 Hungerland bases the denote/signify distinction on this passage, cf. pp. 15, 19, 34, 354, and she seems unsure about the word 'name': she mentions a difference between the Hobbesian and the modern senses, and prefers to talk of 'n-expressions' pp. 22, 51, 55, 62. Martinich too, in commenting on I.ii.3, glosses Hobbes's 'name' as 'word' p. 350.

8 'Nomina signa sunt non rerum sed cogitationum', and 'Nomina quarum rerum sint'. In this part of her argument, Hungerland is following the reluctant lead of J. W. N. Watkins, in his *Hobbes's System of Ideas*, 2nd edn., London, 1973, pp. 101–3.

9 Molesworth, ed., *English Works*, I, p. 17.

10 'Quoniam autem Nomina, ut definitum est, disposita in oratione, signa sunt conceptuum; manifestum est ea non esse signa ipsarum rerum; quo sensu enim intelligi potest sonum hujus vocis *lapis* esse signum *lapidis*, alio quam ut is qui vocem eam audisset colligeret loquentem de lapide cogitasse? Itaque disputatio illa an nomina significent *materiam*, an *formam*, an *compositum*, aliaeque ejusmodi metaphysicorum, errantium sunt, nec intelligentium verba de quibus disputant.'

11 'Nomina per se singula . . . signa . . . non sunt, nisi quatenus in oratione disponuntur.'

12 'Neque vero ut omne nomen, alicujus rei nomen sit, necessarium est. Sicut enim voces *homo*, *arbor*, *lapis*, ipsarum rerum nomina sunt, ita quoque imagines hominis, arboris, lapidis, quae occurrunt somniantibus, sua sibi habent nomina, quamvis res non sint, sed rerum figmenta tantum et phantasmata. Datur enim ipsarum meminisse, ideoque nominibus eas non minus quam res ipsas notari et significari oportet.'

13 See pp. 16–17, 59, 73–102.

14 See pp. 21, 57, 68, 76, 83, 97, 103.

15 '*Homo* quemlibet e multis hominibus, philosophus quemlibet e multis philosophis denotat propter omnium similitudinem.' This is, incidentally, another example of Hobbes using *denotare*, even though on Hungerland's interpretation he is talking about signification.

16 'Propter similitudinem, aequalitatem, vel identitatem.'

17 Page references are to the earliest version of the original 1651 edition (known as the '*Head*'). These are inserted in the text of modern editions such as that of A. R. Waller (Cambridge, 1904), and of C. B. Macpherson (Harmondsworth, 1968).

18 Compare, for example, Locke's remarks in *An Essay Concerning Human Understanding*, II.xi.9, ed. P. H. Nidditch, Oxford, 1975, p. 159. 'Thus the same colour being observed to-day in chalk or snow, which the mind yesterday received from milk, it considers that appearance alone, makes it a representative of all that kind; and having given it the name *whiteness*, it by that sound signifies the same quality wheresoever to be

imagined or met with; and thus universal, whether *ideas* or terms, are made.'

19 Hungerland may have been influenced by Hobbes's remark in I.ii.3, that *signs* would be of no use (*inservire non possent*) to someone who was alone in the world. All Hobbes means is that no-one would bother to use words as signs, unless they thought they had an audience. Hungerland goes much further than this, in insisting that there must be a *real* possibility of communication, see pp. 35–42, within a living linguistic community, see pp. 18, 22, 61, 102. On p. 21 she notes (as Hobbes does not) that, 'The term ['speech'] covers not only a complex of words spoken, but also, mutatis mutandis, a complex of words written'.

20 Even if it could be shown that Hobbes used words like *vox*, *sermo*, *vocare*, and *appellare* in preference to *verbum*, *propositio*, and *denotare* in contexts of actual verbal communication, it would still need to be shown that he avoided such terms in other contexts, such as when referring to language *in general*. But this is not the case: for example, the very titles of his chapters on language in general in *De homine* (ch. x) and *De corpore* (ch. ii) are *De sermone* and *De vocabulis* respectively. Cf. *Leviathan*, ch. iv, 'Of Speech'.

21 'Cum autem philosophiae ut diximus et *notae et signa* (*notae* ut recordari, *signa* ut demonstrare cogitationes nostras valeamus) necessaria sint; nomina utramque rem praestant. Sed *notarum* prius quam *signorum* officio funguntur. Nam homini, etsi unicus in mundo existeret, inservirent quidem ad memoriam, cum tamen ad demonstrationem, nisi alius sit cui demonstret, inservire non possent.'

22 'Natura itaque *nominis* consistit primario in eo quod sit *nota*, adhibita memoriae causa; accidit autem ut serviat quoque significandis, demonstrandisque iis rebus quas memoria tenemus.'

23 'Itaque homo solitarius philosophus fieri sine magistro potest. Adamus potuit. Sed docere, hoc est demonstrare, supponit duos, orationemque syllogisticam.'

24 'Sermo definitus; homini proprius.'

25 'Non . . . verba quatenus verba, sed quatenus signa sunt.'

26 Cf. *De corpore*, I,ii.5

27 Sometimes Hobbes seems to imply that he has in mind any *living* vernacular. For example, when he says: 'Which Terms being meer Words, and in this occasion insignificant, passe onely in Latine, that the vanity of them may bee concealed' (*Leviathan*, ch.xlvi, p. 373). However, it is clear from both the passages quoted below that he allows Latin as used by the Ancient Romans themselves to be a legitimate vernacular for these purposes (though he seems to forget how artificial Cicero's Latin was). Hobbes's target is the 'barbarous' Latin of the Mediaevals and of the Scholastics, as contrasted with the correct and intelligible Latin of the Renaissance Humanists and their followers, such as himself.

28 This second function is of particular importance to philosophers. If a

philosophical position is false, it is not by virtue of being *empirically* false, but because it involves contradiction, incoherence, confusion, misconceptualisation, and other such faults amounting to meaninglessness. To win the argument in a philosophical debate therefore requires a powerful and clear-cut criterion of meaning.

29 The theology that survives is strongly immanent rather than transcendent – as in D. Z. Phillips, *Death and Immortality*, London, 1970, and I. Dilman, *Morality and the Inner Life*, London, 1979, ch. 10. A similar style of interpretation is to be found among late-nineteenth century Hegelians, e.g. B. Bosanquet, *Essays and Addresses*, London, 1889, pp. 92–107, and J. A. Stewart, *The Myths of Plato*, Oxford, 1905, pp. 57–8.

30 Hobbes thus avoids two difficulties which have bedevilled empiricist theories of meaning in particular. First, there is the temptation to dilute the relevant criterion of meaning into vacuity in order to accommodate areas of discourse it would be implausible to reject out of hand as meaningless (for example, Locke's generous use of the notion of an abstract idea, and Berekeley's somewhat less generous use of the notion of a notion). Secondly, some criteria of meaning are so stringent as to outlaw themselves (e.g. some versions of the verification principle), whereas Hobbes's official theory leaves him free to make general, *a priori* claims about meaning itself.

31 Humanism and emergent science were only two movements among many within a complex, shifting alliance of different tendencies opposed to scholasticism. Thus Hobbes is quite exceptional in his hostility towards the various forms of occultism, which were normally part and parcel of both the humanism and the science of the time. See my 'Occultism and Philosophy in the Seventeenth Century' in *Philosophy, its History and Historiography*, ed. A. J. Holland, Dordrecht: Reidel, 1985, pp. 95 – 115.

32 A nice example of the alliance between classicists and scientists against the barbarous and philistine professors of theology is to be found as late as 1699. The Swiss mathematician and experimentalist Johann Bernoulli, at the time a professor at the University of Groningen in Holland, described his theology colleagues to Leibniz in the following terms: 'If you only knew what I have had to put up with here, at the hands of these idiots and ignorant philistines, I am sure I would have your sympathy. It is enough to say that, in order to deem me worthy of their odium, they have branded me as avant-garde, and as the introducer of 'theatrical wisdom' (that is what they insist on calling Experimental Philosophy, which I am the first to teach here). Such is the revulsion of these pathetic little men against the humanities!' (Letter of 3 June, 1699, in C. I. Gerhardt, *Leibnizens mathematische Schriften*, Vol. III, Halle, 1856, p. 590. [My translation].

33 Hobbes mentions French and English as suitably copious. He would not have had to look far for a language which was not copious enough: Leibniz was almost completely defeated by the task of writing philosophy in German, and Kant depended heavily on Latin borrowings. Since then times

have changed: a twentieth-century Hobbes might require that *German* philosophy be translatable into English in order to count as meaningful.

34 Some Wittgensteinians preserve the autonomy of ordinary language through making it non-theoretical by definition. Consequently it is incommensurable with any theoretical stance, and there can be no conflict between, say, a scientific and a commonsense world view, since the latter would have to constitute a rival *theory*. Hobbes's account of how ordinary language becomes contaminated is not worked out in any detail. Thus in the passage from *Leviathan*, ch. viii, p. 39, he says that it is the way words are put together that deprives them of signification; whereas in ch.xlvi, p. 379, he refers to 'insignificant Traines of strange and barbarous words', implying that the words themselves are also meaningless. Again, he says that meaningless discourse is peculiar to those who 'converse in questions of matters incomprehensible', which begs the question of the criterion of meaning.

35 See p. 15 of the 1st Edn. Hobbes's most extensive attack on priests and academics for hoodwinking ordinary folk is in *Leviathan*, ch.xlvii: 'Of the Benefit proceeding from such Darknesse; and to whom it accreweth.'

36 See his 'Anselm's Ontological Arguments', *The Philosophical Review*, 69, 1960, pp. 41–62; reprinted in *The Many-Faced Argument*, eds. J. Hick and A. McGill, London, 1968, pp. 301–320, esp. p. 315. Malcolm's point is that the widespread use of the concept of God as a necessary being is logically sufficient to validate its meaningfulness, and hence to establish the existence of God, as the being whose existence is contingent only upon the meaningfulness of his concept.

37 It is essential not to confuse the private/public distinction with Hobbes's distinction between *notes* and *signs*. There is, of course, a sense in which notes go public as soon as names are used as signs; conversely, signs go private when ordinary language is used for registering thoughts. But this only reinforces the point that the fundamentally private activity of scientific reasoning involves signification as well as notation, and the fundamentally public activity of using ordinary language involves notation as well as signification.

4 *Jan V. Golinski*

Robert Boyle: scepticism and authority in seventeenth-century chemical discourse

Robert Boyle's text of 1661, the *Sceptical Chymist,* enjoys a paradoxical position in the canon of chemical literature, one which raises important historiographical questions.[1] Although it is explicitly sceptical about contemporary claims to chemical knowledge, the work has frequently been assigned a foundational significance for the modern discipline of chemistry. It remains, according to the introduction to the Everyman's Library edition, 'a classic in the annals of science', and at least the first part of Boyle's claim (supposedly embodied in his Irish epitaph) to be 'Father of Chemistry and Uncle of the Earl of Cork' must be supposed to rest on his best-known chemical work.[2] Why then was Boyle, in the text by which his paternity over modern chemistry is conventionally certified, so anxious to disown his own chemical forebears? Why is this supposedly foundational text a work of professed scepticism as to the foundations of chemistry? Why is a work which has attained the status of an authority in the historiography of chemistry in fact avowedly anti-authoritarian?

A straightforward answer to these questions, of a kind which has traditionally been given by historians of chemistry, might run as follows. Boyle was 'sceptical' to the extent that he did not believe in the four 'elements' of Aristotle (earth, air, fire and water), or the three 'principles' of Paracelsus (salt, sulphur and mercury). In the *Sceptical Chymist,* he brought forward experimental evidence to demonstrate the non-existence, or at least the non-producibility, of these 'elements' or 'principles'.[3] In the course of attempting to establish that there were really no such things, he advanced a definition of a chemical element which is close to that recognised today. Although some historians have taken this establishment of the modern definition of an element as a mark of the foundational status of Boyle's text,[4] the more sophisticated have recognised that such an interpretation derives from a misunderstanding of his motives. This kind of answer to the questions I have raised thus suggests an approach to Boyle's text which firmly separates its scepticism from its status as a founding authority in chemical discourse. The scepticism is taken as true to Boyle's intentions, but limited in its application to the elements or principles, while the status of the text as an authority is assigned to the ironies of its subsequent historiographical reputation.

I shall argue that this straightforward answer to my initial questions is profoundly misleading in two senses. Firstly, it limits Boyle's scepticism to the question of the chemical elements or principles, and thereby overlooks its wider application; secondly, it implies that Boyle himself did not intend to establish authoritative foundations for subsequent discourse on chemistry, whereas I believe that he did. My claim is based upon a reading of the *Sceptical Chymist* which exposes the rhetorical intentions of the work. I shall show that Boyle deployed scepticism as a textual strategy, to challenge the authority of a particular discursive tradition. And he did so with the explicit intention of relocating the grounding authority of chemical discourse elsewhere, in the immediate experiences of the chemical laboratory, or in literary narratives which recounted those experiences. For all Boyle's criticisms of the influences of rhetoricity and literary traditions in natural philosophy, his project can be shown to depend on the persuasive power of particular narrative and representational techniques, which were themselves literary in form and rhetorical in intention and in effect.

This reading in turn depends upon my adoption of an historiography which directs attention at the rhetorical dimensions of scientific writing. This historiographical orientation derives from three sources. Firstly, recent work by certain philosophers of language has raised the prestige of the discipline of rhetoric as a key to understanding the operations of language in human communication. Philosophers such as Hans-Georg Gadamer, Jacques Derrida, and Richard Rorty (all of whom have been influenced by Martin Heidegger) have criticised the distinction between the 'content' and 'form' of a text, as is implied in the judgment that the 'content' of a science is a structure of ideas, while its linguistic 'form' is governed simply by the requirements of effective communication of those ideas.[5] By contrast, these philosophers have argued that texts should be assessed as unities, without the imposition of distinctions between linguistic 'form' or 'style' and ideational 'content', and should be analysed in relation to the regularities governing particular modes of discourse directed at particular audiences. Writers such as Gadamer and Roland Barthes have explicitly identified such an attitude to language with that of the classical and Renaissance discipline of rhetoric.[6]

Secondly, philosophically-minded historians of science as diverse in their methods as Michel Foucault and Thomas Kuhn have highlighted the pedagogical and rhetorical importance of scientific texts. Foucault has discussed how academic disciplines and pedagogical practices 'constitute a system of control in the production of discourse'.[7] Kuhn's crucial analytical term 'paradigm' includes among its various meanings that of the embodiment of a scientific achievement in a written text, transmitted by an institutionalised didactic tradition.[8] Taken together, these writers suggest that the historian of science should focus on the rhetorical forms and functions of scientific texts, in relation to the social and institutional context in which science is perpetuated.

Finally, certain historians of science working on the seventeenth-century period have begun to investigate linguistic forms, textual structures, and rhetorical practices of natural philosophy texts. The works of Owen Hannaway, Steven Shapin, Barbara Shapiro, and Simon Schaffer, have been particularly influential in shaping my approach.[9] What follows is intended as a contribution to a new picture which is emerging from their work, of the way in which scientific discourse was reconstructed in the seventeenth century.

In his introduction to a work which was intended as an appendix to the *Sceptical Chymist,* Boyle explained his use of the sceptical mode of argument:

> . . . though sometimes I have occasion to discourse like a Sceptick, yet I am far from being one of that sect; which I take to have been little less prejudicial to natural philosophy than to divinity itself. I do not with the true Scepticks propose doubts to persuade men, that all things are doubtful and will ever remain so, . . . but I propose doubts not only with design, but with hope, of being at length freed from them by the attainment of undoubted truth.[10]

Boyle seems to be saying two important things in this passage.

Firstly, he disavows scepticism as a philosophical doctrine, a belief in the doubtfulness of all knowledge. For Boyle, rather, scepticism is a rhetorical strategy, a mode of discourse to be used, as he says, *on occasion* and *with design.*[11] Secondly, Boyle's 'design' was the achievement of new truths, through calling current beliefs into doubt. As I shall argue, his 'design', in deploying the sceptical strategy of argument, was to challenge the authority of the prevalent mode of chemical discourse, and to initiate a restructured discourse, founded on the persuasive power of narrative accounts of experiments, personally vouchsafed by witnesses.

Towards this end, sceptical arguments provided part of the rhetorical means. Boyle cast the *Sceptical Chymist* in the form of dialogues, in order to communicate his ideas in a didactically-effective way, while also dramatising the rituals of proper philosophical debate. In his view, the manners of such debate required the strict avoidance of dogmatic theorising. Within such a dramatic structure, the deployment of sceptical arguments undercut the authority of traditional chemical discourse, while avoiding the pressure to explicate theoretical alternatives.

Carneades, who speaks for Boyle in the dialogues, adopts the dramatic persona of the sceptic; he repeatedly reminds his colloquators that he is assuming 'the person of a Sceptick'.[12] Playing this role exonerates Carneades from the need to demonstrate his theoretical contentions: 'his Part being chiefly but to propose Doubts and Scruples, he does enough, if he shews that his Adversaries Arguments are not strongly Concluding, though his own be not so neither'. Or, as Carneades himself puts it: ' 'Tis not so much my present Talk to make assertions as to suggest doubts'.[13] Carneades has only to detect failures in the arguments of his colloquators, and can then propose his own hypothesis as a probable explanation of the phenomena. He can even propose inconsistent hypotheses, if each is at least as probable as that being challenged.[14] Operating within these rules, Carneades can suggest, but not establish, matters of theoretical doctrine. And for Boyle, who composed these dramatic dialogues, the voice of the sceptic served as a rhetorical tool, to challenge the authority of certain beliefs, by sowing the seeds of doubt.

On one level, what Boyle was setting out to challenge was the theoretical doctrine of the chemical elements or principles. But on a more general level, his sceptical attack was directed against the authority of traditional chemical discourse as such, of which the doctrine of principles was portrayed as providing the fundamental axioms. Boyle repeatedly insisted that he intended to displace the basic theoretical axioms of the whole discursive tradition of chemistry. Chemists, he wrote, 'deduce things from Salt, Sulphur and Mercury'. The chemical principles were those 'upon which so many others are superstructed', but they were 'false or uncertain principles, to draw . . . consequences from'.[15]

Why did Boyle portray the doctrine of chemical elements or principles as fundamental to the structure of traditional chemical discourse? There are two reasons, I think, one of which has been widely recognised by historians, the other not. Firstly, the doctrine appeared to provide a theory of *qualities* and of *generation*, and it is clear that Boyle intended his own corpuscularian matter-theory to serve the same function.[16] Secondly, by suggesting that the chemical principles provided the axiomatic basis of chemical discourse, Boyle was attacking the systematic *form* of that discourse; and here I think that consideration of the form of chemical texts in the seventeeth century becomes relevant.

The early seventeenth century saw the development, and widespread and growing popularity, of the comprehensive, methodically-arranged, textbook of chemistry. A textual *tradition* was established, which has been traced back to the publication of Andreas Libavius's *Alchemia* in 1597. Following Libavius, the textbooks of Beguin, Brendel and Rolfinck sustained the tradition, wherein aspects of textual organisation, and the verbal forms of discussion of particular subjects, were passed on from author to author.[17] This continuity of subject-matter, and of forms of textual organisation, was complemented by the deployment, in prefaces and on title-pages, of a rhetorical component of the tradition. This rhetoric stressed the comprehensiveness and didactic utility of that text, and generally laid claim to unprecedented clarity of exposition achieved though a more rigorous organisation or 'methodisation' of subject-matter. The key claims, in the rhetorical presentation of chemical textbooks in the seventeenth century, were thus clarity, comprehensiveness, didactic utility, and rigorous textual *method*.[18]

It was these rhetorical claims which Boyle clearly intended to challenge, in propounding a new literary form for chemical discourse. He proposed the practice of writing experimental 'essays', which were to be narrative accounts of laboratory experiments, and justified this by criticising the presentational rhetoric of the textbook tradition.[19] He objected to the systematic treatises, which were produced by adherence to 'the laws of method'. Readers were being deceived by 'the specious and promising titles and comprehensive method of these systems', for example those that 'pretend to give us compleat accounts of chymistry'.[20] Such claims to comprehensiveness, achieved through the use of method, and the deployment of these claims in the titles of works, had been constant features of the seventeeth-century textual tradition. Boyle acknowledged that systematic texts might be educationally useful, but set his own form of chemical discourse against the tradition, by shifting the emphasis from *didactics* to *research*. When writing essays, an author should describe only what he knows from personal experience, so that the description of experimental 'facts' may be clearly demarcated from philosophical or theoretical 'opinions'. A 'perspicuous' style should be used, without ornamentation or foreign terms, and in a 'philosophical [rather] than a rhetorical strain'.[21]

Boyle's scepticism was thus directed at an entire literary tradition, that of the systematic textbooks of the seventeenth century, particularly those on chemistry. That he perceived his target in these terms goes some way towards explaining the persistent denial of his own dependence on any kind of literary tradition, which was a common feature of Boyle's writing. He could, he wrote, 'be very well content to have scarce looked upon any other book than that of nature'.[22] The weakness of his eyes sometimes served as Boyle's excuse for paying closer attention to the 'book of nature' than to books by human authors.[23] He disavowed, on various occasions, detailed knowledge of the works of recent mechanical philosophers or Aristotelians, and he insisted that experimental 'essays' should not contain criticisms of other authors or summaries of their opinions.[24]

Boyle's portrayal of his work as entirely independent of previous natural philosophy, as this had been embodied in a literary tradition, was symptomatic of his general desire to re-make philosophical discourse. The same general aim was responsible for the insistence that all communication, in chemistry and in philosophy generally, should be free of linguistic artifice and rhetorical ornamentation. Initiating a correspondence with Benjamin Worsley, a chemical associate of his in the 1640s, Boyle requested, 'I shall not only consent, but article that not alone all ceremony and compliment, but even all rhetoric and care of language, may be severely banished from our commerce'.[25]

Since he conceived the need to escape from the bonds of determination imposed by the traditional language of chemical discourse, Boyle's scepticism was turned on a linguistic target, for the language of chemistry, as characterised by Boyle, was tortuously obscure, and riddled with ambiguities. He described the usual mode of discourse of chemists as 'their obscure, ambiguous, and almost Ænigmatical Way of expressing what they pretend to teach'.[26] Far from teaching openly in fact, the chemists had used an obscure style accessible only to those who were already '*Sons of Art*'. The most troublesome feature of this obscure style was the ambiguity of key chemical terms. Boyle wrote:

I find that even Eminent Writers . . . do so abuse the termes they employ, that as they will now and then give diverse things, one name; so they will oftentimes give one thing, many Names; and some of them (perhaps) such, as do much more properly signifie some Distinct Body of another kind; nay even in Technical Words or Termes of Art, they refrain not from this Confounding Liberty; but will, as I have Observ'd, call the same Substance, sometimes the Sulphur, and sometimes the Mercury of a Body.[27]

Boyle's sceptical attack on the language of traditional chemical discourse must be seen in relation to two distinct contexts. The first was the debate on language, emphasised in a number of recent historical studies for its close association with developments in natural philosophy in seventeenth-century England.[28] The political crisis in England in the middle years of the century was widely perceived as a crisis of meaning and of language.[29] Proposals for reform included those for the construction of 'universal' or 'philosophical' languages, suggested by George Dalgarno and Francis Lodwick among others.[30] Artificial languages such as these sought to assign to each species of things an unequivocal name, which would define the characteristics of that species in true and unambiguous distinction from all others. Boyle's complaint that the application of names to species of things was largely arbitrary, 'for want of a true characteristic or discriminating notes', echoed the rationale for such artificial languages.[31]

Following the foundation of the Royal Society in 1660, leading members of the Society, such as Robert Hooke and John Wilkins, continued to espouse such schemes.[32] The Society became identified with an attitude to the use of natural language which shared some of the aims of the proponents of philosophical languages; this attitude was expressed most prominently by the Society's historian and apologist Thomas Sprat. Sprat attacked 'luxurity and redundance of *speech*', focusing his censure on 'this vicious abundance of *Phrase*, this trick of *Metaphors*, this volubility of *Tongue*, which makes so great a noise in the World'.[33] Even members of the Society who did not hold to the ambitious schemes for artificial languages were very likely to agree with Sprat on the desirability of restricting the use of natural language to its purely denotative functions, and hence of eliminating obscurity, ambiguity, and the unnecessary elaborations of formal rhetoric.

Furthermore, Boyle was not alone in perceiving that these attitudes had particular implications for *chemical* discourse. Sprat also singled out the 'chymists' for particular criticism, for the 'darkness' of their written style, and he identified what he saw as a significant paradox of recent chemical discourse:

> This desire of glory, and to be counted *Authors*; prevails on all, even on many of the dark and reserv'd *Chymists* themselves: who are ever printing their greatest mysteries; though indeed they seem to dó it with so much reluctancy, and with a willingness to hide still; which makes their *style* to resemble the *smoak*, in which they deal.[34]

Boyle's sceptical disavowal of the language of traditional chemical discourse thus drew encouragement from the attitudes to language of his colleagues in the Royal Society, in the same way that his pretense to be independent of the literary tradition of previous philosophical writing was echoed in the Society's motto: 'Nullius in Verba'. The beliefs on language shared by members of the early Royal Society form one specific context in which we can locate Boyle's sceptical approach to the chemical discursive tradition.

There is, however, another context which should not be overlooked. Boyle was responding to a situation more specific to chemistry, a situation in which communication was coloured by the operations of a *market*, in which applicable information (especially that concerned with medicine and pharmacy) was traded on a commercial basis. His antipathy to the operations of a market in chemical information was made possible by his social position, as a financially-independent, although untitled, younger son of an Anglo-Irish aristocratic family. He wrote, 'Being a bachelor, and through God's bounty furnished with a competant estate for a younger brother, . . . I had no need to pursue lucriferous [wealth-producing] experiments, to which I much preferred luciferous [illuminating] ones'.[35] This social position, vis-à-vis those who engaged in the communication of chemical information as a trade, affected Boyle's attitude to chemical discourse in at least three identifiable ways.

Firstly, Boyle believed that a free and unhindered communication of chemical information could only be achieved if it was not valued as a commodity. This was the theme of his first published work, the tract, 'An invitation to a free and generous communication of secrets and receits in physick', published in 1655, in a volume edited by the Puritan reformer Samuel Hartlib.[36] Here Boyle argued that Christian charity and the welfare of the public both demanded the freest possible exchange of chemical and pharmaceutical information, a practice which was incompatible with the retention of 'secrets and receipts' for commercial gain. These sentiments were similar to those Boyle ascribed to the 'Invisible College', a group with which he became involved in 1646–47, and were close to the ideals of Hartlib whose group he encountered in 1647.[37]

Secondly, Boyle ascribed his lack of commitment to established chemical *theory* to the nature of his social relations with chemical artisans and tradesmen, from whom he had learned chemical operations. He wrote,

> I who had the good Fortune to Learn the Operations from illiterate Persons, upon whose credit I was not Tempted to take up any opinion about them, . . . should be more dispos'd to accommodate the Phænomena that occur'd to me to other Notions than to those of the Spagyrists.[38]

Thirdly, Boyle's distaste for the argumentative style and self-serving dialectical sophistries of chemical language owed something to a gentlemanly disdain for the cut-and-thrust of the market competition between chemical texts. A smooth and civil language was said to be suitable to the *Sceptical Chymist,* 'a book written by a Gentleman, and wherein only Gentlemen are introduced as speakers'.[39] 'A man may be a Champion for Truth, without being an Enemy to Civility', Boyle insisted, and accordingly the characters in the dialogues exchanged elaborate expressions of courtesy.[40] They also, being 'Gentlemen, and very far from the litigious humour of loving to wrangle about words or terms or notions as empty', adopted without dispute the definition of an element which set the terms of the debate.[41] In this way, ambiguous chemical terms could be removed by employing settled definitions, arrived at through gentlemanly consensus.[42] Boyle's scepticism in the face of the inherent obscurities and ambiguities of chemical discourse was, in short, a product of his social position as much as of the general attitudes to the nature of language which he shared with his friends and contemporaries.

Boyle's sceptical attack on chemistry as a literary tradition was only one aspect of his attempt to reform chemical discourse. His criticisms of the linguistic and textual practices of the chemical tradition were complemented by his development of new linguistic and textual practices, designed to re-establish chemical discourse with renewed authority and persuasive power.

According to Boyle, the supreme persuasive power resided in personal experience of laboratory experiments. Experiments, personally witnessed, possessed both a 'novelty or prettiness' which could attract men, and the power to persuade them.[43] They thus achieved greater conviction than that wrought by the rhetorical tricks and dialectical sophistries of the scholastics, which Boyle dismissed as 'being like the tricks of jugglers, whereby men doubt not but they are cheated, though oftentimes they cannot declare by what flights they are imposed on'.[44] Indeed, according to Boyle, Aristotelian natural philosophers had consistently failed to recognise the persuasive power of experiments and had restricted their use to the most basic level of education: 'to satisfy those that are not capable of a nobler conviction . . . they employ experiments rather to illustrate than to confirm their doctrines, as astronomers use sphæres of pasteboard, to descend to the capacities of such as must be taught by their senses'.[45]

The 'chymists', on the other hand, had perceived the demonstrative force of experiments and hence had made much use of them. Their failure had been their inability to develop an appropriate linguistic and literary apparatus, to convey the persuasive power of their experiments in discursive form. This, as Boyle saw it, was the most serious consequence of the chemists' ambiguous language, their failure properly to clarify and define technical terms.

In trying to avoid the complementary errors of Aristotelians and 'chymists', Boyle was forced to engage one of the most urgent problems facing seventeenth-century natural philosophers. As Barbara Shapiro has argued, despite the attacks on rhetoric by Boyle and other members of the early Royal Society, their project remained implicitly rhetorical.[46] The new natural philosophy had still to concern itself with the problem of procuring belief through the persuasive use of language. The categories of evidence, probability and testimony were originally rhetorical and legal categories, which were now made appropriate to the establishment of claims to knowledge. Only the nature of the testimony had changed, the experience of the senses now being more highly valued than literary authority. But even so, the use and assessment of written reports of experiments required some criteria for the conveying and evaluation of literary authority. Boyle therefore set out to develop the rhetorical apparatus which would enable him to convey the persuasive power of experiments in written form.

Boyle recognised that it was true that 'he that hath seen [an experiment] hath more reason to believe it, than he that hath not', but hoped that, if the literary apparatus for the description of experiments could be sufficiently developed, those who had not seen them could be persuaded almost as readily as those who had.[47] This was the goal for which Boyle was aiming, in the successive improvements and refinements in his discursive practices. And he embarked on this with a full awareness of the difficulties involved: 'there is great difference', he wrote, 'betwixt the being able to make experiments, and the being able to give a philosophical account of them'.[48] The rhetorical apparatus which Boyle developed to convey a 'philosophical account' of experiments provided the tools for constructing his experimental essays, and (more generally) suggested the ground-rules for a new order of chemical discourse. In contrast to the dominant forms of the discursive tradition which he criticised, Boyle's experimental essays were *narrative,* as opposed to exhaustively methodical; *suggestive,* rather than conclusively demonstrative; and *subjective,* in the sense that they included personal and circumstantial details which would not be thought relevant by many subsequent scientific writers.

The importance of these personal and circumstantial details in Boyle's discursive practice has recently been emphasised in a stimulating paper by Steven Shapin, describing what he has called Boyle's 'literary technology'.[49] Shapin stresses the importance of the presence of witnesses in establishing the credentials of experimental 'facts', and connects this with Boyle's understanding of the laboratory as a forum for the witnessing of experimental demonstrations by an audience. Shapin portrays Boyle's literary apparatus as designed to produce 'virtual witnessing' of experiments by those who read the descriptive accounts, but who were not actually present at the experimental demonstration. Boyle's literary devices can thus be understood as a series of mechanisms to ensure credibility: prolixity of circumstantial information, detailed illustrations, and what Shapin calls 'appropriate moral postures' (candidness about failed experiments, modesty in advancing philosophical opinions, a reluctance to engage in disputation, etc.).

I have profited considerably from reading Shapin's paper, and would particularly want to reiterate the importance of witnessing, as a means of certifying the factual status of experimental reports. In reporting his experiments, Boyle sometimes listed the names of the witnesses who were present, stressing that they were 'judicious and illustrious'.[50] For example, the experiments on mercury, which he published in the *Philosophical Transactions* of 1675, were said to have been witnessed by Henry Oldenburg, and by 'the Noble and Judicious' Lord Brouncker.[51] Boyle clearly believed that, by giving the names of witnesses of good character and/or high social standing to the experiments he described, he was buttressing the persuasive power of his narrative. In this way, the rhetorical efficacy of Boyle's accounts was made to depend upon their character as testimony, and hence, to a crucial extent, upon the credibility of the witnesses to the events described.

In proposing that experimental reports should be confirmed by identified witnesses, Boyle was attempting to validate a new form of natural-philosophical practice by appealing to the legitimating analogy of legal procedure. Legal procedure was sanctioned by its longevity, and by widespread social prestige, and it was itself being adapted to a new concept of the probability of verdicts, and to new problems of assessment of the credibility of testimony.[52] Boyle saw knowledge as producible in legal, religious and scientific realms by the same process of the mobilisation of the testimony of credible witnesses. In all three fields, 'moral certainty' might be achieved if the testimonies concurred and the witnesses were sincere, knowledgeable and disinterested.[53]

The same criteria which Boyle had established to provide an authoritative basis for a reformed chemical discourse were applied in his appraisal of the legacy of the textual tradition in chemistry. In particular, Boyle's attitude to any chemical text turned upon his assessment of the credibility of the witnesses to the experiments it described. Aside from descriptions of experiments, the remainder of the existing textual tradition was ruled out of court. This included both systematic expositions of theory, and 'receipts' or instructions for preparing useful products. Boyle warned his readers

> not to be forward to believe chymical experiments when they are set down only by way of prescriptions, and not of relations; that is, unless he that delivers them mentions his doing it upon his own proper knowledge, or upon the relation of some credible person, avowing it upon his own experience.[54]

What Boyle was proposing was a wholesale reinterpretation and reappraisal of the chemical discursive tradition. Chemical texts were to be read as a series of narrative 'relations', descriptive accounts of experiments, whose validity turned upon the testimony of the experimenter, or of 'some credible person'. The continuity of the discursive tradition was thus to depend upon the reliability of personal testimony; and it was thus essential that authors were always cited by name. Boyle therefore criticised 'that Indefinite Way of Vouching the Chymists say this, or the Chymists affirme that . . .'.[55]

Boyle dramatised the process of reappraisal of the chemical discursive tradition, in the dialogues of the *Sceptical Chymist*. The pervasive atmosphere was that of a judicial tribunal. Carneades, the sceptic, assumed the mantle of an examining magistrate, announcing his intention to examine 'only the experiments of my adversaries, not their speculative reasons'.[56] The experiments were brought forward, each attested to by witnesses, whose credibility was made the measure of the acceptability of their testimony.

Boyle's own experimental 'relations' were of course accepted without question. They were introduced into the dialogue by Carneades, who mentioned Boyle in the third person as the experimenter and author of the narratives concerned. The testimony of other witnesses was treated rather more sceptically, but it was usually accepted when Carneades (presumably speaking for Boyle) gave assurances as to the eminence or credibility of the witnesses concerned. For example, he claimed that 'two friends of mine, the one a physician, and the other a mathematician, and both of them persons of unsuspected credit', had solemnly assured him that they had performed the reduction of mercury to water.[57] Again, Carneades said that the dissolution of pearls 'has been confirmed to me, upon his own observation, by a person of very great veracity'.[58]

Similarly, the testimony was brought forward of 'a person of very great credit', on the production of vitriol, and of 'a very skilful and credible person' on the gold found in certain ores.[59]

Such assurances of the personal 'credit' of witnesses were essential to Boyle's purposes. His attempt to persuade his readers of the validity of experimental 'facts' depended ultimately upon these assurances of the credibility of the witnesses. Judgments of personal credit formed the foundations for Boyle's new order of chemical discourse. Yet, if not exactly built on sand, Boyle's new edifice of chemical discourse rested on a category of 'credit' which was much less than rock-solid. Crucial problems arose in the reappraisal of the legacies of particular writers in the chemical discursive tradition.

The works of Jan Baptista van Helmont provide an important example. Boyle was sympathetic towards many of Helmont's experiments, especially those which indicated that water might be the single fundamental principle of matter, even though he thought many of his 'relations' were obscure. The problem was whether 'you will credit Helmont'.[60] In general, Boyle tended to look for confirmatory testimony from other witnesses than Helmont himself. 'I think it somewhat harsh to give him the lye, especially as to what he delivers upon his own proper tryal', he wrote, and asserted that he had heard testimony 'from very credible eye-witnesses', and had witnessed experiments himself, which tended to bear Helmont out.[61] Boyle concluded that Helmont was

> an author more considerable for his experiments than many learned men are pleased to think him . . . [But that] so paradoxical a truth as that which these experiments seem to hold forth, need to be confirmed by more witnesses than one, especially since the extravagancies and untruths to be met with in Helmont's treatise of the Magnetic Cure of Wounds, have made his testimonies suspect in his other writings.[62]

Boyle's remarks on Helmont indicate how, in practice, the category of personal 'credit' was separable into two parts, the sincerity of a witness and the plausibility of what he described. To be charitable in ascribing sincerity to other authors was, to Boyle, an elementary rule of the manners of philosophical discourse. It was a matter of 'modesty' to suspect 'some mistake in the manner of the observation, [rather] than that the experimenters did not sincerely deliver it'.[63] Boyle presumably felt that experimental discourse would simply be impossible unless other authors were assumed sincere. But even this assumption was sometimes unjustified in chemistry. In a pessimistic mood, Boyle confided, 'I have so often met with chymical preparations which I have found insincere, that I dare scarce trust any', and as we have seen Helmont's 'credit' as a witness was severely reduced by his description of the impossible phenomenon of the 'magnetic cure of wounds'.[64]

More frequently, Boyle assumed that the author of a particular experimental narrative *was* sincere, even when he failed to replicate the experiments reported. The problem of the sources of the disparity in these cases was what he addressed in his 'Two essays concerning the unsuccessfulness of experiments' of 1661. Boyle asked, why do attempts to replicate experiments sometimes fail, although based on narrative accounts 'communicated for true by candid authors or undistrusted eye witnesses'?[65] Among the answers to this question which he discussed, and particularly relevant in the case of chemical experiments, was variability in the composition of chemical materials. To this factor, together with unknown variations in the circumstances of the experiments, Boyle ascribed his failure to replicate certain reported procedures for manufacturing gold, which he was inclined to believe, although he could not repeat them.[66]

But there was a further reason why the testimonies of sincere eye-witnesses were not always reliable in chemistry. As Boyle pointed out, and as other chemists of the time also noted, eye-witnesses to chemical demonstrations could be deceived.[67] Chemistry had attracted a certain number of 'mountebanks' and 'charlatans', notorious for the display, before a gullible audience, of a fake procedure for the manufacture of gold. In the face of experimental 'demonstrations' which were really conjuring tricks, a proposal to found chemical discourse on eye-witness accounts of experiments was fraught with difficulties. Boyle made the analogy with stories of witchcraft: narratives of this kind could carry conviction, but only if related with the fullest account of the circumstances, and certified by appropriate 'testimonies and authorities'.[68] In both chemistry and the study of witchcraft, a sceptical approach to the assessment of received narratives was an essential complement to the attempt to base experimental discourse on eye-witness testimony.

Boyle's new order of chemical discourse was intended to be a co-operative enterprise. Rather than writing systematic treatises on all aspects of the subject, chemists were to contribute experimental 'facts' towards the eventual compilation of a 'natural history' of experimental phenomena. As Shapin points out, Boyle invited chemists to membership of the 'invisible college' of literary intercourse and correspondence, but made such membership subject to their communicating attested experimental 'facts', written in Boyle's own narrative style.[69] The mediator of communications between Boyle and Continental chemists was often the first Secretary of the Royal Society, Henry Oldenburg. In a letter to the chemist Augustin Boutens, Oldenburg indicated the form which such communication should take:

> the English admire chemical operations undertaken by men of sense who are free from the vulgar prejudices imposed on the world by some people who undertake to speak dogmatically without such a preliminary critical examination as our excellent Mr. Boyle found necessary in his *Sceptical Chymist*.[70]

Oldenburg had perceived the two sides to Boyle's enterprise: the sceptical challenge to the discursive tradition, and the initiation of a new order of chemical discourse founded on narrative descriptions of experimental 'facts'. He portrayed Boyle's *Sceptical Chymist* as a 'critical examination' of the legacy of the discursive tradition, which aimed to free chemists from the impositions of dogmatic discourse. If its lessons were taken to heart, chemical discourse would consist of the exchange of accounts of 'chemical operations' between 'men of sense'. It was this kind of exchange of information which Oldenburg invited from his Continental correspondents.

The responses he received sometimes went beyond simply conveying accounts of operations, raising on occasion quite fundamental problems concerning Boyle's discursive style. For example, the response of Adrien Auzout, to Boyle's narrative descriptions of his experiments, indicated that those descriptions were not perceived by everyone as straightforward and unproblematic as Boyle intended they should be.[71] Auzout wrote that he took Boyle to have been 'presenting his result historically', but was troubled by the fact that 'nearly always some mishap prevented their being carried to a conclusion'. Boyle's candidness over his mistakes and failures in performing experiments had in this case *hindered* the acceptance of his narratives, because it recalled to Auzout the escape-clauses and failed promises of some self-proclaimed gold-makers. This was one possible weakness of the method of description of experiments which relied for its persuasiveness on the personal credibility of the author. Auzout did however note, that as regards Boyle, he was 'certain of the author's sincerity'.[72]

Rather more serious problems were raised by Boyle's discussion of the relationship between chemical experiments and philosophical theory. Boyle did not conceal his partiality for a theory of matter which he called 'corpuscularianism', which combined aspects of Epicurean atomism and Cartesian mechanical philosophy.[73] But he carefully distinguished such philosophical 'opinions' from experimental 'facts', while envisaging a relationship between them in which chemical experiments were to be made 'serviceable' or 'useful to philosophy', by deploying them to 'illustrate' or 'confirm' the corpuscularian matter-theory.[74] For Boyle, this posed the problem of forging discursive connections between experimental narratives and accounts of theory. How was the relationship between the two to be represented rhetorically and textually?

Boyle solved the problem by rejecting a dogmatic presentation of theory and instead maintaining a restrained and consistent probabilism in his claims for the mechanical philosophy. The theory was taken adequately to explain chemical experiments, perhaps even to be confirmed by them, but Boyle never claimed that a rigorous demonstration of the truth of any particular mechanism had been achieved.[75] His arguments in support of the mechanical philosophy were primarily those of the simplicity of its explanatory principles, the general range of their applicability to phenomena, and their 'intelligibility'.[76] It was thus sufficient to show that the experiments could be explained adequately by mechanical principles, without claiming a necessary demonstration.

An example of Boyle's method of presenting theory together with descriptions of chemical experiments is his essay on the analysis and synthesis of saltpetre, in his *Certain Physiological Essays* (1661).[77] In this essay, Boyle first presents an account of his experiment, 'set down nakedly, as I first tried it, by way of narrative'.[78] In the course of the narratives, he makes scrupulous efforts to ensure the credibility of his account, giving practical advice on how to replicate the experiments, satisfying the anticipated doubts of his readers (whom he addresses as an imaginary pupil, Pyrophilus), and candidly confessing his errors and omissions. Having given this narrative of his experiments, Boyle proceeds to a series of 'reflections' or 'opinions', where he rationalises the probable (but no more than probable) mechanisms to account for them. Here he reflects:

This experiment seems to afford us an instance, by which we may discern, that motion, figure, and disposition of parts, and such like primary and mechanical affections . . . of matter, may suffice to produce those more secondary affections of bodies, which are wont to be called sensible qualities.[79]

Boyle's essay on saltpetre was discussed by Benedict Spinoza, in correspondence with Oldenburg in 1662–3.[80] In responding to the essay, Spinoza drew attention to the problem of the connection (cognitive and discursive) between experimental narrative and philosophical conclusions. As he saw it, an alternative, and (in his opinion) more thoroughly mechanistic, theoretical account was equally consistent with the experimental evidence Boyle had presented. The focus was thus directed at Boyle's presentation of the inference from experiment to theory, which Spinoza found severely lacking in rigour. Indeed, as far as Spinoza was concerned, Boyle had committed the mistake of transgressing the boundaries of discursive categories, in even attempting to draw philosophical conclusions from a narrative account of experiments: 'Since Mr. Boyle does not put forward his proofs as mathematical, there will be no need to enquire whether they are altogether convincing'.[81] As for Boyle's attempt to disprove experimentally the Aristotelian theory of substantial forms, Spinoza believed this was entirely misguided. He wrote, 'I do not know why the illustrious Mr. Boyle is so anxious to draw this conclusion from his experiment: for the same has been proved sufficiently (and more than sufficiently) by Bacon, and afterwards by Descartes'. Spinoza's objection was not against Boyle's philosophical conclusions as such, but against their discursive representation as arising directly from a narrative of experiments.

Similarly, when G. W. von Leibniz referred to Boyle's work in his correspondence with Oldenburg, he focused on Boyle's failure to represent his theoretical conclusions on chemistry in the recognised literary form. He twice asked Oldenburg to urge Boyle to write a systematic treatise on chemistry.[82] Notwithstanding Boyle's prolific publications on chemistry, Leibniz wrote, 'I beg you to urge him some time vehemently at least to write distinctly and openly what his opinions on that subject are'.[83] As Leibniz saw it, Boyle had provided some hints to a chemical theory, but his experiments would be wasted unless they were systematised in a properly-structured text.

Only thus could chemistry become 'philosophical', and Boyle could thus benefit the human race by producing 'some body of chemistry'.[84]

As the remarks of Leibniz and Spinoza suggest, the problem of relating experimental narrative to discussion of philosophical theory could be seen as an aspect of the problem of ordering phenomena in philosophical discourse. Among those whose expectations had been formed by the conventions of the seventeeth-century chemical textual tradition, there was a tendency to see the problems of chemical discourse entirely in terms of a lack of the correct methodical order in chemical texts. Boyle, on the other hand, was largely resistant to the temptation to solve the problems of chemical discourse by attempting a more rigorous order in the structure of texts. He declined to follow the model of Francis Bacon's 'method', in which philosophical theory was sought through a process of systematic ordering of phenomena. Discussing his practice of essay-writing, Boyle wrote that he had originally intended to present descriptions of experiments on the model of Bacon's *Sylva Sylvarum*, but he had subsequently decided to abandon 'that succinct way of writing', so as to explain himself at greater length.[85] Because Boyle's essays were essentially narrative accounts of a particular set of experiments, dependent on prolixity of circumstantial detail to convey his personal experience, they did not lend themselves to the rigorous ordering of atomised data which Bacon proposed. Nor did they lay claim to the achievement of philosophical conclusions through a process of methodisation.

Among chemical writers after Boyle, particularly in the early- and mid-eighteenth century, one finds both a renewed interest in Bacon's work, and a renewed emphasis on *order* in chemical discourse, as a means of achieving chemical theory. It was these developments which largely determined the repute in which Boyle's work came to be held. Among eighteenth-century writers, Boyle's sceptical attitude to previous chemical discourse was widely regarded as valuable, but was rarely perceived as an attack on systematic discourse as such. In addition, his narrative accounts of experiments were praised for their clarity, but were not seen as an adequate basis for a 'philosophical' or theoretical chemistry. The latter would only arise from a procedure of methodical tabulation and ordering of chemical phenomena.

The Scot, William Cullen, lecturer in chemistry at Glasgow and

Edinburgh Universities in the middle decades of the eighteenth century, exemplifies these attitudes. In his negative appraisal of traditional textual forms, and his criticisms of the organisational methods they adopted, Cullen remained 'sceptical' in something like Boyle's sense. But he nevertheless displayed a faith in the possibilities of method, particularly the Baconian method of induction, and through method, in the possibility of attaining a 'philosophical chemistry'. Cullen's divergence from Boyle in this respect dictated his assessment of Boyle's achievement. He praised him for having cleared chemistry of 'that jargon & affected mystery which had hitherto deformed it', and for having 'delivered his experiments with plainness, perspicuity & Candour', but he could not concede that Boyle had rigorously established a mechanistic chemical theory, which he took to have been his intention.[86]

Thus Cullen limited Boyle's sceptical appraisal of preceding chemical discourse to the reform of vocabulary, and overlooked or undervalued his attempts to produce a new chemical discourse, based on the rhetorical force of personally-witnessed experimental 'facts'. Neither the foundational nor the sceptical side of Boyle's enterprise in the *Sceptical Chymist* appear to have been properly appreciated one century after its publication. The pattern of interpretation of Boyle's work had been established which would give it its present paradoxical position, as a celebrated but misunderstood classic of scientific literature.

I have argued in this paper that the misunderstandings which have surrounded the *Sceptical Chymist* derive from a failure to apprehend the work's rhetorical dimensions. The example of Cullen illustrates how the requirements of teaching caused chemical discourse to revert to a structure based on the systematic ordering of phenomena and the overt explication of theory. In this context, the radicalism of Boyle's intentions was not perceived, because the discursive structure of the free exchange of experimental narratives, which he had espoused, could not be sustained. As historians, however, we cannot do Boyle's work justice unless we recognise the radicalism of his rhetorical intentions, unless we perceive that his scepticism was directed against the entire structure of seventeeth-century chemical discourse, and that his intentions for reform embraced no less than the complete restructuring of that discourse. In this way, a reading of the *Sceptical Chymist* which is alert to its rhetorical dimension can restore the work to its proper place in the historiography of chemistry, and perhaps can also provide a model to be applied to other crucial texts of the period of the Scientific Revolution.

Notes

1 Robert Boyle, *The Sceptical Chymist: Or Chymico-Physical Doubts & Paradoxes, Touching the Spagyrists' Principles Commonly call'd Hypostatical,* London, 1661; facsimile reproduction, London, 1965. Referred to hereafter as Boyle, *Sceptical Chymist.*

2 Robert Boyle, *The Sceptical Chymist,* London, 1911; reprinted, 1964, p. xii. Boyle's Irish epitaph is reported in W. C. Dampier, *A History of Science and its Relations with Philosophy and Religion,* Cambridge, 1971, p. 141.

3 Marie Boas (also known as Marie Boas Hall), *Robert Boyle and Seventeenth-Century Chemistry,* Cambridge, 1958, pp. 75–141.

4 See, for example, J. R. Partington, *A History of Chemistry,* 4 vols., London, 1961–4, 1970, II, p. 496.

5 See for example, M. Heidegger, *On the Way to Language,* tr. P. D. Hertz, San Francisco, 1971; H.-G. Gadamer, *Philosophical Hermeneutics,* tr. D. E. Linge, Berkeley, 1976; J. Derrida, *Writing and Difference,* tr. A. Bass, London, 1978; R. Rorty: *Philosophy and the Mirror of Nature,* Princeton, 1979; 'Philosophy as a kind of writing', *New Literary History,* X, 1978–9, pp. 141–60.

6 Gadamer, *Philosophical Hermeneutics,* pp. 21–6; R. Barthes, 'Science versus literature', *Times Literary Supplement,* 28 September, 1967, pp. 897–8.

7 M. Foucault, 'The discourse on language', in his *The Archaeology of Knowledge and the Discourse on Language,* tr. A. M. Sheridan Smith, New York, 1976, p. 224.

8 T. S. Kuhn, *The Structure of Scientific Revolutions,* Chicago, 1962.

9 References to these works are given in the notes which follow. A more recent study, which attempts to apply a similar approach to the 'Chemical Revolution' of the late-eighteenth century, is Wilda C. Anderson, *Between the Library and the Laboratory: the Language of Chemistry in Eighteenth-Century France,* Baltimore and London, 1984.

10 'Experiments and notes about the producibleness of *chymical principles,* being parts of an appendix designed to be added to the *Sceptical Chymist',* in R. Boyle, *The Works of the Honourable Robert Boyle,* ed. T. Birch, 6 vols., 2nd edn., London, 1772, I, pp. 587–661 (quotation on p. 591).

11 Compare the study of scepticism as a philosophical doctrine in the early-modern period, by Richard H. Popkin, *The History of Scepticism from Erasmus to Spinoza,* Berkeley and Los Angeles, 1979.

12 Boyle, *Sceptical Chymist,* pp. 47, 185.

13 *Ibid.*, sig. A4r, pp. 379–80.
14 *Ibid.*, sig. A4v, p. 12.
15 *Ibid.*, sig. A2v, p. 10; Boyle, *Works*, I, p. 587.
16 See Boas, *Robert Boyle*; T. S. Kuhn, 'Robert Boyle and structural chemistry in the seventeenth century', *Isis*, XLIII, 1952, pp. 12–36; Hélène Metzger, *Les Doctrines Chimiques en France du Début du XVIIe à la Fin du XVIIIe Siècle*, Première partie, Paris, 1923, pp. 251–66.
17 On the seventeenth-century chemical textual tradition, see Owen Hannaway, *The Chemists and the Word: The Didactic Origins of Chemistry*, Baltimore and London, 1975, especially pp. 152–6; J. V. Golinski, 'Language, Method and Theory in British Chemical Discourse, c.1660–1770', University of Leeds Ph.D. thesis, 1984; J. R. R. Christie and J. V. Golinski, 'The spreading of the word: new directions in the historiography of chemistry 1600–1800', *History of Science*, XX, 1982, pp. 235–64.
18 Compare the similar situation with natural philosophy texts, as described in Patricia Reif, 'The textbook tradition in natural philosophy 1600–1650', *Journal of the History of Ideas*, XXX, 1969, pp. 17–32.
19 'A proëmial essay, wherein, with some considerations touching experimental essays in general, is interwoven such an introduction to all those written by the author, as is necessary to be perused for the better understanding of them', Boyle, *Works*, I, pp. 299–318.
20 *Ibid.*, I, pp. 300–301.
21 *Ibid.*, I, pp. 304–5.
22 *Ibid.*, I, p. 317.
23 *Ibid.*; cf. I, p. 229; III, p. 9.
24 *Ibid.*; pp. 311–12, 355–6.
25 Boyle to [Benjamin Worsley], [c. February 1647], Boyle, *Works*, VI, pp. 39–40. For discussion of the correspondent and dating of this letter, see Charles Webster, *The Great Instauration: Science, Medicine and Reform, 1626–1660*, London, 1975, pp. 59–60 and fnn.
26 Boyle, *Sceptical Chymist*, sig. A3r.
27 *Ibid.*, pp. 200–201.
28 Mary M. Slaughter, *Universal Languages and Scientific Taxonomy in the Seventeenth Century*, Cambridge, 1982; Barbara Shapiro, *Probability and Certainty in Seventeenth-Century England*, Princeton, 1983, pp. 227–66.
29 That it was so perceived by Hobbes, for example, is argued in Miriam M. Reik, *The Golden Lands of Thomas Hobbes*, Detroit, 1977, pp. 42–3, 81–112.
30 Slaughter, *Universal Languages*, pp. 116–19 (on Lodwick), 141–53 (on Dalgarno); Vivian Salmon, ed., *The Works of Francis Lodwick*, London, 1972.
31 Boyle, *Works*, III, p. 50.
32 Slaughter, *Universal Languages*, pp. 157–186 (on Wilkins and

Hooke); Hans Aarsleff, 'Wilkins, John', *Dictionary of Scientific Biography,* XV, pp. 361–81.

33 Thomas Sprat, *The History of the Royal Society of London for the Improving of Natural Knowledge,* London, 1667; reprinted, St. Louis, 1958, pp. 111–13.

34 *Ibid.,* pp. 37, 74.

35 Boyle, *Works,* I, p. cxxx.

36 [Robert Boyle], 'An invitation to a free and generous communication of secrets and receits in physick', in Samuel Hartlib, ed., *Chymical, Medicinal and Chyrurgical Addresses: Made to Samuel Hartlib, Esquire,* London, 1655, pp. 113–50. See also Margaret Rowbottom, 'The earliest published writing of Robert Boyle', *Annals of Science,* VI, 1948–50, pp. 376–89; R. E. W. Maddison, 'The earliest published writing of Robert Boyle', *Annals of Science,* XVII, 1961, pp. 165–73. Cf. the context in which the tract is interpreted by James R. Jacob, *Robert Boyle and the English Revolution,* New York, 1977, pp. 32–5.

37 See Boyle to Francis Tallents, 20 February, 1647, in Boyle, *Works,* I, pp. xxxiv–xxxv; also II, p. 123; IV, p. 150; and for background, Webster, *Great Instauration,* pp. 57–67; R. E. W. Maddison, *The Life of the Honourable Robert Boyle, F.R.S.,* London, 1969, pp. 67–8. Compare Boyle's ideals with the more explicitly Millenarian expectations of chemists such as John French and William Cooper: J. French, *The Art of Distillation,* London, 1653, sig. B4r; W. Cooper, *The Philosophical Epitaph of W.C. Esquire,* London, 1673, 'Author's Epistle' at the beginning of books two and three (unnumbered pp.).

38 Boyle, *Sceptical Chymist,* sigs. A7v–A8r. (By the 'Spagyrists', Boyle meant the Paracelsian chemists.)

39 *Ibid.,* sig. A6r.

40 *Ibid.,* sig. A6r, pp. 1–9, 14.

41 *Ibid.,* p. 16.

42 As Shapin and Schaffer have argued, such processes of the engineering of consensus among gentlemanly discussants found expression in the aims and procedures of the early Royal Society. See Steven Shapin, 'Pump and circumstance: Robert Boyle's literary technology', *Social Studies of Science,* XIV, 1984, pp. 481–520; Simon Schaffer, 'Making certain', *ibid.,* XIV, 1984, pp. 137–52; and cf. Sprat, *History,* pp. 91, 98–102.

43 Boyle, *Works,* III, p. 5.

44 Boyle, *Sceptical Chymist,* p. 15.

45 *Ibid.,* pp. 20–21.

46 Shapiro, *Probability and Certainty,* pp. 3–14.

47 Boyle, *Sceptical Chymist,* p. 78.

48 *Ibid.,* p. 208.

49 Shapin, 'Pump and circumstance'.

50 Robert G. Frank, *Harvey and the Oxford Physiologists,* Berkeley,

Los Angeles and London, 1980, p. 132.

51 B. J. T. Dobbs, *The Foundations of Newton's Alchemy,* Cambridge, 1975, p. 194.

52 Shapiro, *Probability and Certainty,* ch. 5; Schaffer, 'Making certain'.

53 On the legal analogy, see Boyle, *Works,* IV, p. 182; on religion, see V, p. 529.

54 Boyle, *Sceptical Chymist,* sig. A3r.

55 *Ibid.,* sig. A3v.

56 *Ibid.,* p. 295.

57 *Ibid.,* p. 134.

58 *Ibid.,* p. 280.

59 *Ibid.,* pp. 365, 371.

60 *Ibid.,* p. 231.

61 *Ibid.,* p. 78.

62 *Ibid.,* p. 113.

63 Boyle, *Works,* I, p. 152.

64 *Ibid.,* I, p. 320.

65 *Ibid.,* I, pp. 318–53 (quotation on p. 318).

66 *Ibid.,* I, pp. 320ff., 334–5.

67 Boyle, *Sceptical Chymist,* pp. 181–85. For other examples of reports of mountebanks, see *The Works of the Highly Experienced Chymist John Rudolph Glauber,* 3 vols. in 1, tr. Christopher Packe, London, 1689, I, p. 221; Walter Harris, *Pharmacologia Anti-Empirica: Or a Rational Discourse of Remedies both Chymical and Galenical,* London, 1683, pp. 298–332.

68 Boyle to Joseph Glanvill, 18 September, 1677, in Boyle, *Works,* VI, pp. 57–9, esp. p. 58. See also [Robert Boyle], *An Historical Account of a Degradation of Gold Made by an Anti-Elixir,* London, 1678, p. 8.

69 Shapin, 'Pump and circumstance', pp. 16–17, 26–7.

70 Oldenburg to Boutens, 11 November, 1667, in A. R. Hall and M. B. Hall, eds., *The Correspondence of Henry Oldenburg,* 11 vols. to date, vols. I–IX, Maddison and Milwaukee, 1965–73; vols. X, XI, London, 1975–7, III, p. 590.

71 Auzout to Oldenburg, 18 December, 1666, *Ibid.,* III, pp. 291–9.

72 *Ibid.,* III, p. 296.

73 See, for example, *Experiments, Notes &c. about the Mechanical Origin or Production of divers particular Qualities,* Boyle, *Works,* IV, pp. 230–353; *Some Specimens of an Attempt to make Chymical Experiments useful to illustrate the Notions of the Corpuscular Philosophy, Ibid.,* I, pp. 354–9. On Boyle and the mechanical philosophy in general, see Frank, *Harvey,* pp. 93–7; J. R. Jacob, 'The ideological origins of Robert Boyle's natural philosophy', *Journal of European Studies,* II, 1972, pp. 1–21; *idem., Robert Boyle*; Robert Kargon, *Atomism in England from Hariot to Newton,* Oxford, 1966; *idem.,* 'Walter Charleton, Robert Boyle and the

acceptance of Epicurean atomism in England', *Isis*, LV, 1964, pp. 184–92.

74 Boyle, *Works*, I, pp. 354–5, 358, 359.

75 *Ibid.*, IV, pp. 232, 235; see also Kargon, *Atomism*, pp. 102–4.

76 Boyle, *Works*, I, pp. 355, 589; IV, pp. 277, 281.

77 *A Physico-Chymical Essay, containing an Experiment, with some Considerations touching the differing Parts and Redintegration of Salt-Petre*, Boyle, *Works*, I, pp. 359–76.

78 *Ibid.*, I, p. 359.

79 *Ibid.*, I, p. 364.

80 April, 1662 – August, 1663, in Hall and Hall, eds., *Oldenburg Correspondence*, I, pp. 448–70; II, pp. 37–43, 86–96, 101–4.

81 *Ibid.*, I, p. 462.

82 *Ibid.*, II, 42–7, 301–7.

83 *Ibid.*, II, p. 306.

84 *Ibid.*, II, p. 46.

85 Boyle, *Works*, I, pp. 305–6.

86 Cullen Papers, box 2, item 44, Glasgow University Library.

5 *Tony Davies*

The Ark in flames: science, language and education in seventeenth-century England

On the night of the fourth and fifth of September, 1666 the great fire of London, moving rapidly westwards before a strong prevailing wind, circled back north and east of St Paul's and engulfed the parish church of St Lawrence Jewry and the surrounding houses, including the vicarage. Among the incumbent's possessions consumed in the holocaust was the manuscript, with some printed pages in proof, of a large book, the fruit of many years' work by several minds and hands; and in amongst the voluminous drafts, notes and diagrams as they blackened and vaporised in the incandescent air could have been found a computation, combining biblical scholarship, Anglican piety and the strictest scientific method, of the dimensions and internal volume of the Ark, the number and species of animals contained in it, and the amount of food, animal as well as vegetable, required to sustain them during a voyage of exactly three hundred and sixty-five days. Thus, among the unregarded casualties of the great fire, which devastated two thirds of the City and left 200,000 Londoners homeless, must be reckoned, on paper at least, 137 herbivores, forty carnivores, twenty-four 'Beasts feeding on Fruits, Roots and Insects', and 1,825 sheep. Not to mention 109,500 solid cubits of dried hay. Plus, of course, the Patriarch himself, accompanied by his family – the second progenitors of humankind.

The mathematical vicar of St Lawrence Jewry was John Wilkins, Doctor of Divinity, brother-in-law of the late Oliver Cromwell, Fellow of the recently-created Royal Society and one-time Warden of Wadham College, Oxford. And the lost manuscript, which he rewrote with astonishing rapidity and published only two years later, was the first version of his *Essay Towards a Real Character and a Philosophical Language*. Wilkins, who succeeded to the Bishopric of Chester in the same year (1668) as the publication of this, his most celebrated work, is well worth a monograph on his own account. Certainly any investigation of the sociological character of the English Revolution that aims to take account in a non-reductive way of its intellectual and cultural as well as its political and economic features must pay some attention to him. He was the son of an Oxford goldsmith, from a family both puritan and interested in the natural sciences, or 'experimental philosophy'. Wilkins *père* had a noted preoccupation with perpetual motion; and his son's first book, published when he was twenty-four, set out to demonstrate that the moon was a habitable world, as well as – in a later edition – that it could be reached by 'volitation'. In London in the later forties he was one of the leading members of a group of scientific intellectuals and political radicals who aimed to give some practical substance to Bacon's utopian 'House of Salomon', instituted, in Bacon's words, to promote

> the knowledge of Causes, and Secrett Motions of Things: and the Enlarging of the bounds of Humane Empire, to the Effecting of all Things Possible.[1]

In the parliamentary reform of the University of Oxford, undertaken in 1648 in a spirit of Baconian 'advancement' and involving the wholesale ejection of existing fellows and their replacement by organic intellectuals of the revolutionary regime, Wilkins was appointed Warden of Wadham and, as co-Chancellor of the University alongside the more thoroughgoing root-and-branch independent John Owen, played an important part in its political as well as its academic life. During the fifties the Wilkins group commuted between Oxford and London, and the original founding members like Ralph Bathurst and the mathematician John Wallis were joined by younger students of Wilkins such as Seth Ward, Thomas Sprat and Christopher Wren. In 1659 he was elected Master of Trinity College, Cambridge; and though he lost that post at the Restoration he negotiated the political transition adroitly and became, three years later, one of the founding Fellows of the Royal Society. By inclination Presbyterian, he avoided open rupture with the established Church, working for a 'composition' or reconciliation between Dissenters and Anglicans. Praised as a moderate and abused as a trimmer, at once biblical fundamentalist and experimental scientist,[2] Cromwellian functionary and pillar of the ecclesiastical establishment, he seems typical to the point of allegory, a characteristically 'bourgeois' figure in that postrevolutionary 'period of regret' which, according to Marx, capitalist society must undergo 'until it has learnt to assimilate soberly the achievements of its period of storm and stress'.[3] And in nothing is he more typical than in his sober passion for 'enumeration'.

5.1 *Language and number*

For Thomas Hobbes, all thought is an arithmetical, indeed a *commercial* transaction, involving the computation and exchange of 'parcels' or finite quantities of practical consciousness, tabulated and communicated through the stable, conventional currency of language.

> When a man *Reasoneth,* hee does nothing else but conceive a summe totall, from *Addition* of parcels; or conceive a Remainder, from *Substraction* of one summe from another . . . REASON, in this sense, is nothing but *Reckoning* (that is, Adding and Substracting) of the Consequences of generall names agreed upon, for the *marking* and *signifying* of our thoughts.[4]

Linguistic modernism in the seventeenth century has two related motivations. One is formal and epistemological, seeking to reduce the complex historical actuality of language to a system of fixed laws and relationships, and pointing to the emergence in the period of that general epistemic project that Foucault has called 'a universal science of order.' The other is moral and political, and is concerned with the public nature of meaning and the legislation of semantic discipline. In both respects it marks a sharp break with Renaissance thinking about language. For example, the Elizabethan educationalist Richard Mulcaster, arguing against the use of classical grammars as a model for the teaching of English, had insisted on the mutable and profoundly unsystematic character of the vernacular, a character resulting from the intrinsic operations of its 'soulish substance', 'secret misterie' and 'quikning spirit'.[5] In this view, the defining feature of a healthy vernacular is its capacity to generate metaphor and to tolerate neologism. But for Hobbes this conception of language as an ever-moving stream, a medium of innovation, creativity and semantic indeterminacy, is quite simply meaningless, itself a phantasmagoria of those 'metaphors, and senseless and ambiguous words' that he likened to '*ignes fatui*', adding that 'reasoning upon them, is wandring amongst innumerable absurdities; and their end, contention, and sedition, or contempt'. Indeed, Mulcaster's quasi-scholastic metaphors of 'soulish substance' and 'quikning spirit' seem directly to invite *Leviathan*'s celebrated denunciation of 'insignificant sounds':

when men make a name of two Names, whose significations are contradictory and inconsistent; as this name, an *incorporeall body,* or (which is all one) an *incorporeall substance* . . . the words *In-powred vertue, In-blown vertue,* are as absurd and insignificant, as a *round quadrangle.*[6]

Hobbes's stylistic brutalism is all his own; or rather, it belongs to an earlier, more 'Jacobean' mode of pungently reductive aphorism, recalling Bacon or Jonson. But his desire to purge language of 'ambiguity', to expel metaphor and outlaw neologistic 'phrasing', to eradicate all traces of what Vološinov called its 'multiaccentuality' and to reduce it to a rational system of invariant and univalent signs, was shared by the reforming intellectuals of the Wilkins group. Sprat deplored 'this vicious abundance of phrase, this trick of metaphors', and was inclined to believe that 'eloquence ought to be banished out of all civil societies, as a thing fatal to peace and good manners';[7] and Wilkins counterposed the simplicity and functional clarity of his invented 'philosophical [i.e. scientific] language' against the 'ambiguity of words by reason of *Metaphor* and *Phraseology*' found in all 'instituted languages', features which

prejudice the native simplicity of [speech], and contribute to the disguising of it with false appearances . . . witness the present Age, especially the late times, wherein this grand imposture of Phrases hath almost eaten out solid Knowledge in all professions.[8]

Here the double philosophical-political motivation can be seen in the guarded reference to the 'late times', whose studied imprecision permits contrary decodings, as a reference to the Commonwealth or to the Clarendonian reaction of the early sixties. And although Wilkins has been described as a 'moderate', it is worth noting that no distinction of any value can be drawn, at the level of discourse, between the 'liberalism' of his proposals and the supposed absolutism, even 'totalitarianism', of the Hobbesian scheme. On the contrary, while Hobbes only *grumbles* about the semantic treacherousness of natural language and dreams of a world as clean and single-minded as a theorem of Euclid, Wilkins aspires, in his sensible pragmatic fashion, to abolish language altogether and to replace it with a '*Real universal Character,* that should not signifie *words,* but *things* and *notions*', thus inverting the traditional relation between speech and writing, and making spoken language – the 'philosophical language' of the title – into a second-order phonological system wholly subordinated to a fixed and codified *script* – the 'real character'.

This undertaking deserves to be recognised as one of the great seventeenth-century utopias, not so much for its linguistic project, in which Wilkins had been anticipated and indeed actively assisted by Urquhart's 'universal language' and Dalgarno's 'universal character and new rational language',[9] as for the recognition, not shared by those predecessors, that the linguistic enterprise would necessitate a vastly greater one: the 'just enumeration' and systematic classification of *everything in the universe,* from the Creator to the '*cacilla* or slow worm',[10] along with all their parts, attributes and relations. It is proper to call this utopian, not only because of its scale and comprehensiveness, but because at its heart lies an authentically utopian concept: the abolition of the distinction between words and things. Like many of his contemporaries, Wilkins was impressed by reports of the ideogrammatic written language of China. He believed, too, that the Hebrew alphabet had, in its primitive simplicity, encoded a clear and unmediated relation to the world, then in its freshly-created prime. But the first of these was too elaborate, the second too corrupted by subsequent variation and use; and in any case, unencumbered though they might be with the confusion of linguistic diversity and uncontaminated by the gross subjectivity of speech, they remained only *codes,* conventional or pictorial representations of a still separated universe of things. The ontological estrangement between words and things, the primal curse of referentiality through which all sin and error had entered the world, would be resolved only if some means could be devised 'to represent things by such peculiar *signs* and *names* as should express their *natures*' – a language not, as for Hobbes, conventional but authentically *natural* or 'real', and so capable not only of eliminating all misunderstanding and contention, civil, religious and scientific, but of revoking the punishment of Babel itself:

> for supposing such a thing as is here proposed, could be well established, it would be the surest remedy that could be against the Curse of Confusion, by rendering all other *Languages* and *Characters* useless.[11]

All this may sound like, and has been taken for, an epistemological realism, subordinating language to the preexistent and inexorable forms of the actual. In this vein, Wilkins has been seen as a representative, in an admittedly extreme and eccentric version, of the familiar seventeenth-century argument for 'plain style' – for what his friend Sprat called 'the mathematical plainness', involving a return 'back to the primitive purity, and shortness, when men delivered so many things, almost in an equal number of words';[12] and, through that, as typifying the ideological interests of a mercantile petty bourgeoisie. In this historiographic tradition, which has its *marxisant* as well as its whiggish forms and which typically looks to the seventeenth century for forerunners and anticipations of later and more definite developments, the urbane, cautious, politically noncommittal middle-class puritan intellectuals or '*virtuosi*' of the Royal Society, who modelled their discourse, in Sprat's phrase, on 'the language of artizans, countrymen and merchants', have been claimed as precursors of the commercial and utilitarian materialism of the eighteenth and nineteenth centuries. But there is a danger here of writing history backwards, or in circles: not only of reading retrospectively from effects to presumed causes, but of finding the essential features of the former already inscribed, in teleological embryo, in the latter.

Against this tendency to subsume the Royal Society *virtuosi* into a not-yet-fully-developed bourgeoisie, and their interests and values into a not-yet-quite-explicit materialism and realism, it needs to be said that in spite of his conventionally Baconian assertion that '*things* are better than *words*', it is not nature that is 'real' in Wilkins' book but his invented language, and that so far from submitting himself to the experimental discipline of things, he imperiously subordinates them, indeed transforms them, to *writing*. In that exemplary computation of the contents of the Ark, what is at stake is not the historical plausibility of scripture, still less the truth of God's (conventionally *spoken*) word, but the validity of a taxonomic and enumerative system whose only conceivable mode of existence is in writing. For Wilkins, the universe is an issue not for natural history, still less for theology, but for grammatology. If the ancient opposition of word and thing is to be resolved back into a primitive unity, it will be done, not by acknowledging the necessary primacy of things, but by enforcing within the 'artificiall world' of the book the unchallengeable sovereignty of the printed word: 'annihilating all that's made', not to a thought, but to an enumeration, a list of signs and numbers on a page.

5.2 *Education and the new order*

'Before being its object', Jacques Derrida has written, 'writing is the condition of the episteme';[13] and it is certainly true that a preoccupation with the problematics of language appears in the seventeeth century less as an issue in its own right – a linguistics – than as a recurrent theme across a whole range of discourses: legal, political, philosophical, educational. For lawyers, for instance, the determination to establish an authentic tradition of English Common Law brought with it a sharpened sense of the political implications of 'pedler's French and Law-Latin'. A Leveller petition of 1647 dubbed it 'a Badg [sic] of our Slavery to a Norman Conqueror, to have our Laws in the French Tongue, and little less than brutish vassalage to be bound to walk by Laws which the People cannot know', demanding in consequence that

> all the Laws and Customes of this Realm be immediately written in our Mothers Tongue without any abreviations of words, and the most known vulgar hand . . . and that Writs, Processes, and Enroulments, be issued forth, entered or inrouled in English, and such manner of writing as aforesaid.[14]

In tune with this and many similar solicitations, the Rump Parliament passed in 1650 an Act requiring all existing law books to be translated into English, and stipulating that in future, on pain of a fine of twenty pounds,

> all Patents, Commissions, Records, Judgements, Statutes, Recognizances, Rolls, Entries, and Proceedings of Courts Leet, Courts Baron, and Customary Courts, and all Proceedings whatsoever in any Courts of Justice within this Commonwealth, and which concerns the Law, and Administration of Justice, shall be in the English Tongue onely, and not in Latin or French, or any other Language than English . . . And that the same, and every of them, shall be written in an ordinary, usual and legible Hand and Character, and not in any Hand commonly called Court-hand.[15]

Here, as in other instances of vernacularisation in the period – in Culpeper's medical textbooks, for example, or Eachard's attack on pulpit-Latin – a cultural traditionalism, which in one variant saw 'the plaine true English tongue' not merely as the language of the sturdy, independent Saxons but as the veritable speech of Eden itself, joins hands with a pragmatic modernism whose stress falls on ease and clarity of communication and the immediate utility of the communicated knowledge. These criteria were widely current, and served as common ground – or at least as a set of what Christopher Hill has called 'overlapping circles'[16] – for a variety of social groupings and interests, from itinerant antinomian sectaries expecting the imminent delivery – in English – of the Last Judgement to army agitators, Leveller intellectuals, and solidly respectable bourgeois republicans like Harrington and Milton. But the centre of this predicative discourse, the point at which it receives the busiest thematic elaboration and assumes the most urgent practical force, is its intrication with the debate about useful knowledge and the 'advancement of learning': the transformation of the institutions, practices and aims of public education.

Elizabethan educators had, naturally enough, been interested in questions of language; and some, like Mulcaster, had been strong promoters of vernacular schooling. But the early preoccupations were largely formal and philological, concerned as they were to bring English out of the shadow of Latin and to establish native regularities of grammar, orthography and lexis. Spenser's *Shepherd's Calendar,* for example, was praised for its attempt 'to restore, as to theyr rightfull heritage such good and naturall English words, as haue ben long time out of vse and almost cleane disherited'.[17] But although the notion of a linguistic 'heritage', and still more the implications of a language both 'good and naturall', seem to adumbrate much later and more developed ideologies of 'Englishness', the connections are not explicitly made, and the central definitions remain lexical and syntactic: 'for the knitting of sentences . . . what in most English wryters vseth to be loose, and as it were vngyrt, in this Authour is well grounded, finely framed, and strongly trussed vp together'. The crucial shift to semantics, and through that to a conception of language as a *systemic* bearer of ideological meanings and values, along with an asssociated shift from speech to writing as the paradigm of language use, occurs in the revolutionary decades. And although an antiquarian strain persists, it serves in the forties and fifties only to throw into relief the conscious, indeed apocalyptic and utopian modernism of the educational project.

Milton's is the most widely-known statement of this project:

The end then of learning is to repair the ruins of our first parents by regaining to know God aright, and out of that knowledge to love him, to imitate him, to be like him, as we may the nearest by possessing our souls of true virtue . . . But because our understanding cannot in this body found itself but on sensible things, nor arrive so clearly to the knowledge of God and things invisible, as by orderly conning over the visible and inferior creature, the same method is necessarily to be followed in all discreet teaching.[18]

But although Milton pays formal acknowledgement to the idea that 'language is but the instrument conveying to us things useful to be known', his retention, within a still fairly orthodox theology, of the traditional conception of the hierarchy of knowledge, in which the study of 'sensible things' adumbrates and symbolises 'knowledge of God and things invisible', points in a different direction, and distinguishes him from the secular and pragmatic radicalism of, say, John Dury:

The true end of all Humane learning is to supply in ourselves and other [sic] the defects, which proceed from our Ignorance of the nature and use of the Creatures . . . nothing is to be accounted a matter of true learning amongst men, which is not directly serviceable unto Mankind towards the supply of some of these defects which deprive us of some part of our natural happiness.[19]

Dury's *Reformed School,* like Milton's *Tractate,* was written in response to the solicitations of Samuel Hartlib, whose energetic activity in the cause of educational reform resulted in a brief visit to London, in 1641, by the very eminent Moravian philologist and 'pansophian' educationalist John Amos Comenius.[20] Hartlib and his associates were seeking to attract public money and to enlist the support of the Long Parliament for a thoroughgoing reformation of public education, including, in an early version, the establishment of a 'probationary college' (a kind of teacher-training college) and a university in London. Nothing came of these projects; but it is worth noting, in view of the cranky and impractical utopianism that is often ascribed to them, that this preoccupation with the institutions appropriate to a reformed national education, with questions of structure and administration, is highly characteristic of the Hartlib group, expressing a studious practicality evident everywhere in their writings, but suggestive too of a tendency to view education as a matter less of pedagogy and curriculum than of administrative and institutional arrangements. Thus George Snell, in another contribution to the Hartlib discussions, propounded a 'designe of schooling' in which a centralised national education system, consisting of vernacular 'rural colleges' in every town and city teaching 'several degrees of schooling, and . . . nothing but that which is useful', would be superintended by a 'Corporation of Schoolmasters' and administered by a Parliamentary Standing Committee of State. This remarkable anticipation was elaborated in a paper to Hartlib in which, signing himself 'Syndicus Reipublicae Literariae' (Officer of the Republic of Letters), Snell spelt out also the ideological purpose of his scheme of 'moral and intellectual reform', calling for

> the advancement of prudential literature, shewing how all sortes of youth nobel, or of inferiour qualitie, rich or poore, may be well educated in their own townes and shires, in all kindes of literarie knowledge, that may most fittlie prepare them to know, love, and obey, the laws and ordinancies of theire owne countrie.[21]

Dury, too, regretted the ignorance among the young of 'the constitution and government of their own country', as of 'the common forms of transaction between man and man', and related it to the more fundamental absence of a centralised and systematic vernacular education in terms which, by associating writing with both drawing and counting, seek to situate it as at once an iconographic and an enumerative activity, a picture of the world and a computation of its contents:

> Children are not taught to pronounce and write accurately in their own language . . . are not taught to draw pictures when being taught to write . . . are not taught rudiments of arithmetic at the same time as writing.[22]

At this point, clearly, the 'pansophian' project of the Hartlib group, with its distinctively educational and institutional concerns, comes close to the more abstractly philosophical and experimental interests both of the Wilkins group and of Boyle's slightly later, and similar, 'Invisible College'.[23] As early as 1641 Comenius had proposed both a *Priscianus Vernaculus* (a mother-tongue primer 'to teach boys the structure of human speech and its component parts') and an *Encylopaedia Sensualium* ('in which are to be described in order all the things which they experience through their senses at that tender age'). And Comenius's collaborator and rival, the Silesian scholar Cyprian Kinner, discussed with Hartlib his idea of

> using for botanical description a vocabulary of special words, each consonant, vowel and syllable of which would convey an entire meaning; e.g. the

consonants in the first syllable denoting qualities, the vowels in that syllable the degrees of qualities. Each word would be a name for a particular thing and serve as a mnemonical aid for its description.[24]

In this way, Kinner argued, 'the language of symbols would be accommodated to man's thoughts about things and would be . . . a remedy for the Babylonian confusion of tongues'.

Milton, in *Of Education,* called Hartlib's scheme 'one of the greatest and noblest designs . . . for the want whereof this nation perishes'. That was in 1644. But throughout the forties the Long Parliament, to whom all these plans and exhortations were addressed, had, it must have seemed, more pressing concerns, and neither Hartlib's enthusiasm nor the international prestige of Comenius were able to accomplish much by way of official acknowledgement. On the 20 July, 1653, however, at the end of a long day of political programming, the Nominated or 'Barebones' Parliament – successor to the ejected Rump and very likely the most radical national assembly ever seen in England, then or since – did finally establish a Standing Committee for the Advancement of Learning.[25] And though that Parliament, which was racked by internal contradictions and fiercely attacked for the character of its legislative agenda (which included, in addition to education, the abolition of tithes and the reform of the legal profession), had been dissolved in its turn before the year was out, the Council of State kept the project alive, even financing its investigations from the proceeds of confiscated estates in Ireland. The reactionary Anthony Wood saw the Committee as a plot 'to suppress Universities and all Schools for Learning, as heathenish and unnecessary'.[26] But he needn't have worried. By 1657 something like twelve thousand pounds had been accumulated; but William Petty, veteran of Boyle's Invisible College, political arithmetician and now Deputy Surveyor-General for Ireland, who had his own notions about how to promote the advancement of learning, obstructed the release of the money and so frustrated the scheme.[27] In 1659 Hartlib made a last-minute appeal to Wilkins, now at Trinity, to lend his prestige and respectability to a projected 'Agency for the Advancement of Universal Learning'. But by now Wilkins too was otherwise engaged. And in any case, it was already too late.

5.3 *Writing*

Drawing attention to the central role occupied by language in Gramsci's account of the formation of nation states and national bourgeoisies, Colin Mercer has remarked recently that 'it is interesting to consider the (largely ignored) importance attributed to language by the intellectual 'fathers' of . . . the bourgeois democracies', and has suggested that it reflects a concern with 'the generation of *common meanings,* literally common *sense* for a population organised in distinctively new ways'.[28]

There are questions of historiography here. The political revolution of the mid-seventeenth century has been exhaustively written and rewritten; the intellectual and cultural revolution, no less interesting and momentous, has received less attention, often being reduced to a kind of second-order 'history of ideas', a bit of 'seventeenth-century background'. So Wilkins and his associates enjoy at best a sort of respectable vicarious obscurity in the prescientific twilight before Boyle and Newton, accompanied by a condescending acknowledgement of their primitivism, their engaging crankiness. And as for Hartlib and Dury, they share their dusty footnote in the history of education: earnest, humourless, frozen in the comic pathos of historical redundancy. Only Christopher Hill and Charles Webster have recognised the interest and significance of these groupings, and attempted to understand them as parts of a single complex social and intellectual phenomenon, an organic intelligentsia characterised by a shared though internally differentiated 'Baconianism', and connected through an intricate nexus of educational, professional and family relationships. Webster's interest, as a historian of science, is primarily epistemological and methodological; but he is not blind to the sociology of scientific modernism, noting how in the fifties, and still to a substantial degree in the reactionary sixties, 'adherence to the experimental philosophy became one of the most acceptable qualifications for intellectuals wishing to seek a career in the universities or in the civil service'.[29] This remained, of course, uncodified – a very English kind of establishment, regulated not by specified procedures of certification and promotion but by an informal network of influence and dependency: Pepys's familiar round of theatre, sermon and dinner-party, in a milieu of gossip and intrigue. It remained so even after the Wilkins group and the Invisible College had together acquired the more palpable embodiment – as the Royal Society – of an official charter and a quasi-collegiate structure, a development that served to consolidate the professional and ideological ascendancy, after 1660, of the 'conservative Baconians', and the concomitant marginalisation of those radicals, like Hartlib and Snell, who had looked for the creation of an entirely new system of educational institutions, and who drew their models, not from the conspiratorial sodality of a college of fellows, but from the unvarnished declarative openness of an anabaptist conventicle.

Christopher Hill, though, reminds us that such distinctions, like the related oppositions of 'mathematicians' and 'experimentalists', 'discoverers' and 'utilitarians', can mislead, disguising deeper and more complex connections and differentiations.[30] Taking issue with H. F. Kearney's traditionalist assertion that the early Fellows of the Royal Society, 'serious and professional' in contrast to the raffish amateur enthusiasts of the Hartlib circle, 'defended the universities and traditional learning against the attacks of the puritan pamphleteers', Hill has stressed instead the continuing 'utilitarianism' of Boyle and Wilkins, the 'professionalism' of Hartlib and Petty, and has argued that Sprat's *History,* accepted at something like face-value by Kearney and other conservative historians of science, was in fact prudentially designed 'to cover up the Parliamentarian and puritan background of the leading founders of the Society'.[31] Other historians have recently demonstrated a convergence of scientific interest between Wallis, Wren, Boyle and Hartlib, as late as 1659–60, around the work of the German chemist and alchemical adept Peter Staehl,[32] and have argued that far from embodying a narrow and politically obsequious 'professionalism', the early Royal Society contained a substantial number of 'social visionaries' who 'looked forward to the emergence of a society integrated in specific ways calculated to serve specific interests'.[33]

To call those interests 'bourgeois' may not be to say very much; but a fairly representative list of those early Fellows, such as Bathurst, Boyle, Petty, Sprat, Ward, Wilkins and Wren, includes only one – Boyle – who did not come from a provincial and identifiably commercial or petty-professional background: two sons of country parsons, one of a Hampshire clothier, one of an Oxford goldsmith, one of a Hertfordshire attorney, and so on. And apart from Boyle, the privately-educated fourteenth child of an Irish peerage, all but one (Christopher Wren, who went to Westminster School) received their schooling, in the words of Sprat's biographer, 'at a little school by the churchyard side'. All, that is, except William Petty, who had little formal schooling of any kind, went to sea in his teens, studied with Hobbes in Paris in the forties, and is reported to have declined a peerage in later life on the impeccably bourgeois grounds that he 'would sooner be a copper farthing of intrinsic value than a brass half-crown, how gaudily soever it be stamped or gilded'.[34]

There are, then, histories here that have yet to be fully written, of the fractions of an emergent class and its associated intellectuals caught in the turbulent commotions of a new political and cultural order. These may well be, as Mercer has suggested, 'Gramscian' histories: monographic accounts of intellectual groupings, philological investigations of the unceasingly negotiated, shifting relations between official ideologies, formal knowledges and popular conceptions of the world, triangulations of the spasmodic and the deeper rhythms of historical continuity and change. But there are other histories too: epistemic soundings, enquiries not so much into cultural meanings and values as into the preconditions of any meaning and value; and one of these – following, carefully, Foucault and Derrida as well as Gramsci – will be a contribution to the history of *writing*.

The Elizabethan grammarians, orthographers and educators, inaugurating the project of a 'ruled' or standardised vernacular, nonetheless regarded speech as the primary form of language, and writing as a second, imitative mode. Writing, in this (quite traditional) view, represents not things, but the mental and phonetic representations of things already present in (spoken) words. In this they remained true to the central humanist concept of *eloquence* (*eloquentia* = speaking out), of the voice as the authentic sign of being: 'speak, that I may see thee'.[35] Social valuations, then as now, underscored the difference. For the 'modern prince' addressed by the discourse of humanist education, to speak well was an essential virtue, writing merely the menial skill of the copyist and the clerical drudge. Lower still in the hierarchy of social value, representing the mechanical objectivity of an indefinitely repeatable writing, without the soul of an authentic voice, was the proto-proletarian activity of the print-shop.

Any account of the intellectual revolution of the mid-seventeenth century is likely to give a prominent place to John Milton; but these modal oppositions provide one way of indicating Milton's real distance, in spite of apparent similarities, from the scientific and educational radicals of the Hartlib group. For his conception of language remains unshakably phonocentric. The 'uncouth swain' of *Lycidas* 'sings' his lament for his dead friend. There is no writing in Paradise, and the creating logos (*legein* = to speak) of *Paradise Lost* is not written but dictated (*dictare* = to speak) to and by the blind poet, to an indifferent series of unseen and largely unremembered scribes. *Areopagitica,* ostensibly a defence of unlicensed printing, is actually a neoclassical oration (*oratio* = speech) addressed to a Parliament (a *speaking* place). And the final, despairing appeal of the *Ready and Easy Way* assumes, climactically, the naked authenticity of a voice, crying in the wilderness.

Against this, the neo-Baconianism of the fifties and sixties, 'radical' and 'conservative' alike, represents an attempted shift of a fundamental kind. When Dury assimilates writing, not to speech, with its dominating presence and its ineradicable social accents, but to drawing and number, as an immediate graphic representation and enumeration of the world-as-it-is, he directly challenges the humanist paradigm. When Wilkins founds his encyclopaedic project, not in a new *language* (a speech-system), but in an invented *character* (a writing-system), he subverts the dominant relations of voice and script, and adumbrates a universe of traces, turning ceaselessly in a Pascalian silence. Derrida has suggested that the grammatological modernism of the seventeenth century remains at heart 'logocentric' because it subordinates its universe of written signs to a prepotent subjectivity, an immanent presence.[36] But this may be to confuse the real force of an argument with the form of its presentation, in which tactical considerations will always play a part. Rather as those contemporaries who called Hobbes an atheist had a surer sense of the likely historical effectivity and discursive drift of his writings than later commentators who have discovered in him the residues of an essential theology, so Wilkins's Ark is real, floats securely in its epistemological element, not because it transcribes or represents something already guaranteed by the authenticity of a creating logos but because it can be written and rewritten; enumerated, depicted, printed. Destroyed in a conflagration in which many thought they heard the voice of an avenging deity, it survived, too – thanks, not to any divine dispensation, but to the secular and mechanical repeatability of writing – the consuming tongues of flame.

The significance of writing in the evolution of bourgeois society needs no emphasis. The historical importance and paradigmatic status of printing; the value attached to contracts, to documentary records, to legal statutes and written constitutions; the often superstitious veneration accorded to the signature (quite literally, the person's 'character'), to the 'very letter of the bond': all these have been extensively noted. In similar and closely related ways, the 'classic realism' of bourgeois aesthetics, the sustained *effets de réalité* of its painting, drama and fiction, all point to a new, if highly equivocal, preoccupation with the exact depiction, transcription and enumeration of the world. And the institutional and taxographic absolutism of Hartlib and Wilkins is seen again, on a more ambitious and determined scale, in the Jacobin enterprise, with its neologistic calendar, its invented state religion, its utopian passion for legal and administrative order.

But such an account is of course misleadingly one-sided. If the insurrectionary conjuncture of the new order can be understood as the moment of *writing*, drawing its energy and new-found legitimacy from the impersonal and collective authority of documents (parliamentary remonstrances, bills of impeachment, sentences of death, instruments of government), the 'organic' elaboration and consolidation of that order, with its oedipal guilts and counter-revolutionary regressions, its tactical reconciliations with the old and its strategic betrayals of the new, will certainly find it necessary to reinterpret, amend or repress those originating texts. And in that hegemonic process, writing may find itself once more modally subordinated to the elusive but authoritative presence of the voice: actual voices, of course – 'BBC' or 'Oxbridge' or 'educated' voices, effortless vocalisations (they like to assume) of grammatical and social orthodoxy; but also, more basically, the phonocentric paradigm itself.

Literature can provide a suggestive instance of this process. Among the inhabitants of Wilkins's grammatological universe may be numbered the protagonists of those eighteenth-century novels that model their narrative and enunciative form, not on oral or dramatic narrations, but on domestic genres of writing – letters, diaries, memoirs. Many of the most notable of these (Moll Flanders, Pamela Andrews, Clarissa Harlowe) are women; and the epistolary mode, in particular, testifies to their exclusion from – indeed, their oppression by – that public and dominative sphere that D. H.

Lawrence (for whom 'utterance' was paradigmatically male) calls 'the spoken world'.[37] Later, too, though the mode itself is largely superseded by more comprehensive and 'omniscient' varieties of realism, the silencing, invisibility and subordination of women continue to be registered in terms suggestive of their dependence on writing. Wentworth's scribbled note in *Persuasion,* the self-tormenting intimacies of Miss Wade in *Little Dorritt,* the fatally misplaced letter in *Tess of the D'Urbervilles*: all stand in the strongest possible contrast to the ebullient masculinity, the picaresque performative vocality of Fielding, Thackeray, Dickens or Lawrence.

This points to the likelihood that, as Terry Eagleton suggests in his account of Richardson's *Clarissa,* the history of writing in the last three hundred years will be in important part the history of sexuality, of cultural representations and enactments of sex and gender.[38] But other kinds of subordination, too, may be condensed or codified into the double articulation of language, as voice and text. Wherever the authoritative voice silences, threatens or excludes, writing, anonymous, collective, unencumbered with affectations of presence and authenticity, may remain an available form of resistance, refusal, difference; perhaps even, at times, a means – through flood and fire – of sheer survival. There is a tradition, among intellectuals, of romanticism about voices: the voice of the oppressed, the *vox populi,* George Orwell's 'voice from the crowd', the recent interest in oral history. This is understandable enough. But perhaps we should be careful before we subscribe wholeheartedly to the power and immediacy of the vocal mode, as a model of communicative practice. The voice of the people was also, to complete the phrase, the 'voice of God'. The voice from the crowd expressed – to Orwell rather admirably – a bored or cynical indifference to politics.[39]

'The science of linguistics', Derrida has written, critically, 'determines language . . . in the last instance and in the irreducible simplicity of its essence, as the unity of the *phonè,* the *glossa* and the *logos*'.[40] Certainly, any cultural theory or historiographic practice that grounds itself, even implicitly, in the concept of the special authenticity of the voice will find it difficult to escape the pull of the current back towards the centrality, conformity and interpellative authoritarianism of which Derrida accuses structuralist linguistics. Perhaps, then, Hartlib, Dury, Wilkins are not so cranky nor so remote from us after all. For they remind us, subjects and exponents of the vocal mode, that writing is valuably and subversively heterogeneous and uncentered, a kind of communicative doodling. It withstands the totalitarian aspirations of the logos, its hunger for plenitude. Wilkins even remembered to leave space in the Ark for the animals that didn't exist, perhaps because they had already become extinct. Writing has no truck with 'nature'. Nobody ever thought to pronounce, as the first term in a philosophical anthropology, that 'man is a writing animal'. It is something we can all learn to do, with and between one another. Those who have no voice in the order of things must become grammatologists, invent a writing. But of course, they have their own voices too; and in time they will be heard.

Notes

An earlier version of this paper appeared in *Literature, Teaching, Politics,* IV, 1985, pp. 63–84.

1 Francis Bacon, *New Atlantis* (*c.*1623), ed. A. B. Gough, Oxford, 1915, p. 35.

2 His *Discourse Concerning a New Planet* as been described by Nicholas Tyack as 'the first full-scale English defence of Copernicanism'. See 'Science and religion at Oxford before the Civil War', in Donald Pennington and Keith Thomas, eds., *Puritans and Revolutionaries,* Oxford, 1978, p. 91.

3 Karl Marx, *The Eighteenth Brumaire of Louis Bonaparte,* in *Surveys from Exile,* Harmondsworth, 1973, p. 150.

4 Thomas Hobbes, *Leviathan* (1651), Harmondsworth, 1968, pp. 100–101.

5 For Mulcaster, see Richard Foster Jones, *The Triumph of the English Language,* London, 1953.

6 Hobbes, *Leviathan,* p. 108.

7 Thomas Sprat, *History of the Royal Society,* London, 1667, pp. 11–13. Sprat called for the foundation of an Academy on the French model (the Académie Française was established in 1635), to regulate the linguistic effects of the Civil War ('which is a time wherein all Languages use, if ever, to increase by extraordinary degree') and to bring the language 'to its last Perfection'.

8 John Wilkins, *An Essay Towards a Real Character and a Philosophical Language,* London, 1668, pp. 17–18.

9 For Urquhart and Dalgarno, see J. E. Sadler, *J. A. Comenius and the Concept of Universal Education,* London, 1966. For the political and economic motivations behind linguistic modernism, see the Royal Proclamation prefixed to Dalgarno's *Ars Signorum,* which describes the universal language as 'of singular use, for facilitating the matter of Communication

and Intercourse between People of different Languages, and consequently a proper and effectual Means, for advancing all the parts of Real and Useful Knowledge, Civilising barbarous Nations, Propagating the Gospel, and encreasing Traffique and Commerce'.

10 Wilkins, *Essay*, pp. 3, 161.

11 *Ibid.*, p. 13.

12 Sprat, *History*, p. 113.

13 Jacques Derrida, *Of Grammatology*, Baltimore, 1976.

14 *A Declaration of Some Proceedings*, 1649. Quoted by Jones, *Triumph*, p. 317, where it is tentatively ascribed to John Lilburne.

15 Statute of 22 November, 1650. See *Acts and Ordinances of the Interregnum 1642–1660*, eds. C. H. Firth and R. S. Rait, 3 vols., London, 1911, II, pp. 455–6.

16 Christopher Hill, *Milton and the English Revolution*, Harmondsworth, 1979, p. 107.

17 Edmund Spenser, *Poetical Works*, eds. J. C. Smith and E. de Selincourt, Oxford, 1912, p. 417.

18 John Milton, *Of Education* (1644), in *Selected Prose*, ed. M. W. Wallace, Oxford, 1925, p. 146.

19 John Dury, *The Reformed School* (*c.*1650), ed. H. M. Knox, Liverpool, 1958, pp. 37–8.

20 See G. H. Turnbull, *Hartlib, Dury and Comenius*, Liverpool, 1947.

21 *Ibid.*, pp. 46–7.

22 *Ibid.*, pp. 50ff.

23 For the Invisible College, see Charles Webster, 'New light on the Invisible College', *Transactions of the Royal Historical Society*, XXIV, 1974, pp. 19–42.

24 Turnbull, *Hartlib*, pp. 439–40. For Comenius's proposals, see p. 360.

25 For the Barebones Committees, see Austin Woolrych, *Commonwealth to Protectorate*, Oxford, 1982.

26 Quoted in Charles Webster, *The Great Instauration: Science, Medicine and Reform 1626–1660*, London, 1975, p. 197.

27 For Petty's proposals, see Turnbull, *Hartlib*, p. 44.

28 Colin Mercer, 'After Gramsci', *Screen Education*, XXXVI, 1980, p. 14.

29 Webster, *Instauration*, p. 519.

30 All these terms are from H. F. Kearney, 'Puritanism, capitalism and the scientific revolution', *Past and Present*, XXVIII, 1964, pp. 81–101.

31 Christopher Hill in *Past and Present*, XXIX, 1965, p. 95. Webster (*Instauration*, p. 54) makes the similar point that Sprat's failure to mention the important scientific seminars convened in London and Oxford in 1645 by the Comenian pansophist Theodore Haak and attended regularly by Wilkins, Wallis, Jonathan Goddard, George Ent and other future Royal

Society *éminences* reflects 'his positive desire to suppress an account of events which might attract attention to the puritan and pansophic associations of the early Society'.

32 K. Theodore Hoppen, 'The nature of the early Royal Society', in W. R. Owens, ed., *Seventeenth Century England: a Changing Culture. Vol. 2, Modern Studies,* London, 1980, pp. 231–40. For Staehl's lectures, see p. 234.

33 J. R. Jacob, 'Restoration, reformation and the origins of the Royal Society', in Owens, *Seventeenth Century England,* pp. 241–52. Jacobs gives as an example of the Fellows' wider political interests Boyle's active involvement in the Council for Foreign Plantations and the Company for the Propagation of the Gospel in New England.

34 The anecdote comes from 'Petty, Sir William', *Dictionary of National Biography,* Vol. 45, pp. 113–19.

35 Ben Jonson, 'Discoveries', in *Complete Poems,* ed. G. Parfitt, Harmondsworth, 1975, p. 435.

36 Derrida, *Grammatology,* pp. 78–9: 'In spite of all the differences that separate the projects of universal language or writings at this time . . . it always leads to an infinitist theology and to the logos or the infinite understanding of God'.

37 D. H. Lawrence, *The Rainbow,* Harmondsworth, 1949, p. 8.

38 Terry Eagleton, *The Rape of Clarissa: Writing, Sexuality and Class Struggle in Samuel Richardson,* Oxford, 1982.

39 George Orwell, 'Inside the Whale', in *Collected Essays,* London, 1961, p. 128.

40 Derrida, *Grammatology,* p. 29.

The perfect cheat: Locke and empiricism's rhetoric

The title of this paper is derived from a famous and much-quoted passage in Book III of Locke's *Essay Concerning Human Understanding,*[1] in which rhetoric is denounced:

> If we would speak of Things as they are, we must allow, that all the art of Rhetorick, besides Order and Clearness, all the artificial and figurative application of Words Eloquence hath invented, are for nothing else but to insinuate wrong *Ideas,* move the Passions, and thereby mislead the Judgment; and so indeed are perfect cheat . . .[2]

The chapter in which this denunciation appears is entitled 'Of the Abuses of Words', and the passage on rhetoric ends that chapter, being the last of the 'abuses' Locke lists. The chapter follows that dealing with the 'Imperfections' of words, and its object is distinguished from such Imperfections in that whereas the latter are, in principle at least, faults of language itself rather than of its users, the former, the abuses, involve what Locke calls '*wilful Faults and Neglects*', of which men are 'guilty'.[3] 'Imperfections' involve problems which are properly epistemological, 'Abuses' problems which are rather moral and political in nature. The chapter on these abuses is followed by one devoted to suggested 'remedies' to both imperfections and abuses, and these remedies attempt to minimise the imperfections and eradicate the abuses through a purification, in the sense of an expurgation or an expulsion. Locke will not attempt to reform language in general, although the reasons for not attempting such a reform are not themsevles so much philosophical as practical and even worldly, based at once on a measured despair with 'Men' and on a disapproval of the 'vanity' such a reform would imply on the part of a reformer, and motivated by the eminently worldly fear of appearing 'ridiculous'.[4] He will try to expel abuse from, and reduce the effects of imperfection in, not the whole of language, but the area of philosophy, the domain of those '*who* pretend *seriously* to *search after,* or maintain *Truth*',[5] or, as the section on rhetoric has it, the place 'where Truth and Knowledge are concerned'.[6] Philosophy should thus be an enclave of clarity within the 'world' (the 'world', as the writings on education tell us, being the place of deceit, desire and the novel), and should establish and maintain frontiers against encroachment from that other space, while restraining itself, with a certain regret, from making forays or crusades into that space. For example, discussing the redoubtable problems posed by the 'Names of *Substances*', Locke says that conformity of these names with things 'as they exist' is 'absolutely necessary in Enquiries after philosophical Knowledge, and in Controversies about Truth', and continues:

> And though it would be well too, if it extended it self to common Conversation, and the ordinary Affairs of Life; yet I think, that is scarce to be expected. Vulgar Notions suit vulgar Discourses: and both, though confused enough, yet serve pretty well the Market, and the Wake. Merchants and Lovers, Cooks and Taylors, have Words wherewithal to dispatch their ordinary Affairs; and so, I think, might Philosophers and Disputants too, if they had a Mind to understand, and to be clearly understood.[7]

Or, in the same vein: 'the Market and Exchange must be left to their own ways of Talking, and Gossippings not be robb'd of their ancient Privilege . . .'[8]

It is worth noting immediately, although for the moment I do no more than note, that this tolerant division of discursive space and the peaceful topology it implies is troubled by Locke's persistent implication that the creation of too closed a space for philosophy can lead to the sterile wrangling of the Schools, and the concomitant desire, voiced in his 'Epistle to the Reader', to make philosophy 'fit' and 'capable' 'to be brought into well-bred Company, and polite Conversation', this 'bringing' of Philosophy 'into' the 'world' going with a certain incursive violence marked in the project to 'break in upon the Sanctuary of Vanity and Ignorance', in the 'Service of Humane Understanding'.

Leaving aside this trouble and violence, it would seem at first that the more benign picture of discursive spaces sketched above is valid in the particular 'abuse' of language constituted by rhetoric. For in discussing this case, Locke admits the existence of discourses 'where we seek rather Pleasure and Delight, than Information and Instruction',[9] and in such discourses would seem to allow, a little grudgingly, the '*figurative Speeches,* and allusion in Language' which he condemns in philosophy. Such uses of language are 'laudable or allowable' (it is worth pondering the value of the 'or' in this formulation), in 'Harangues and popular Addresses', but are 'wholly to be avoided' in 'all Discourses that pretend to inform or instruct', and, further, 'where Truth and Knowledge are concerned, cannot but be thought a great fault, either of the Language or Person that makes use of them'.[10] So rhetoric is, perhaps, acceptable in its place, and that place is not philosophy; there is a frontier to be respected and policed. But, on the other hand, it would seem that despite a show of tolerance, philosophy only constitutes this frontier by transgressing it and by occupying provisionally, as a sort of pre-emptive strike, the ground of rhetoric itself. For the value of the 'laudable or acceptable' compliment condescendingly paid to rhetoric in its place is doubtful, given the prior condemnation of rhetoric's general aim to 'insinuate wrong *Ideas,* move the passions, and thereby mislead the Judgment'. Rhetoric's status as 'perfect cheat' is a *general* fault of rhetoric *as such,* which is thus really deprived of any legitimate ground at all. Its ground is in principle already that of illegitimacy, of planned deceit rather than simply error.

This status of rhetoric can be clarified by comparison with a passage in which Locke is condemning the abuse involved in using words inconsistently in reference to collections of simple ideas;

> Words being intended for signs of my *Ideas,* to make them known to others, not by any natural signification, but by a voluntary imposition, 'tis plain cheat and abuse, when I make them stand sometimes for one thing, and sometimes for another; the wilful doing whereof, can be imputed to nothing but great folly, or greater dishonesty. . . . If Men should do so in their Reckonings, I wonder who would have to do with them? One who would speak thus, in the Affairs and Business of the World, and call 8 sometimes seven, and sometimes nine, as best served his Advantage, would presently have clapp'ed upon him one of the two Names Men constantly are disgusted with. And yet in Arguings, and learned Contests, the same sort of proceeding passes commonly for Wit and Learning: but to me it appears a greater dishonesty, than the misplacing of Counters, in the casting up a Debt; and the cheat the greater, by now much Truth is of greater concernment and value, than money.[11]

Despite obvious analogies, the difference between this type of 'plain cheat' and rhetoric's 'perfect cheat' is perhaps the following. This 'plain cheat', along with other abuses listed in the chapter, is worth denouncing in these terms because it can be remedied by a sort of intellectual and ethical hygiene: the 'Errors and Obscurity', the 'Mistakes and Confusion',[12] and the moral turpitude their use can imply can be avoided by applying the remedies Locke puts forward in Chapter XI. But rhetoric, the 'perfect cheat', is introduced and expelled in this single section, given neither detailed discussion nor detailed remedies. There is no need, says Locke, to list the various 'artificial and figurative application of Words Eloquence hath invented', because that is already the job of rhetoric *itself,* and is thus *already* in the domain of the cheat. Nor is there any need to suggest remedies to rhetoric, as there is in the case of the other imperfections and abuses. Rhetoric is not an imperfection, it is *perfect* cheat; and if it is abuse, it is not a local and corrigible abuse, but a counter-move to Truth and Knowledge *in general.* Further, following an apparently non-philosophical reasoning similar to that used against any project to reform language in general, Locke states that although philosophy can deplore rhetoric, it cannot pretend to eradicate it, because rhetoric, concerned to 'move the Passions', is itself the *object* of a passion, of love. The passion that rhetoric-as-deceit moves is already the passion *for* rhetoric-as-deceit:

> 'Tis evident how much Men love to deceive, and be deceived, since Rhetorick, that powerful instrument of Error and Deceit, has its established Professors, is publickly taught, and has always been had in great Reputation: And, I doubt not, but it will be thought great boldness, if not brutality in me, to have said thus much against it. *Eloquence,* like the fair Sex, has too prevailing Beauties in it, to suffer it self ever to be spoken against. And 'tis vain to find fault with those Arts of Deceiving, wherein Men find pleasure to be Deceived.[13]

The artful, feminine perfection of the 'cheat' of Rhetoric is such as to avoid having 'clapp'd upon' it one of those two unnamed names 'Men constantly are disgusted with'. By avoiding saying something like, ' 'Tis pity she's a whore', Locke is already seduced into his nicely rhetorical final sentence, perhaps because he knew that saying something like ' 'Tis pity she's a whore' would not have saved him either. The brief step, or *pas,* of philosophy into rhetoric in this section introduces a veiled glimpse of the veiled delights of desire; Locke then moves quickly onto the remedies for what can be remedied, finishing his book on language, and gets on with the stern business 'Of Knowledge in General'.

Let me recapitulate briefly: rhetoric is not a local and corrigible abuse of language, a partial threat to Truth and Knowledge, as are the other imperfections and abuses. It involves, not so much a set of blunders or unfortunate inadequacies, nor yet a local dishonesty, as a completely different attitude towards language and its use, an incommensurability which is a generalised threat. Locke can be so brief and perfunctory in his treatment of rhetoric because his common ground with it is merely that of language. But he can be concerned by it and irritated with it because there is that common ground nevertheless. Having moved into the question of language in Book III of the *Essay* rather to his own surprise,[14] Locke inevitably confronts rhetoric somewhere. The rapidity and force with which the problem is discussed and dispatched, and the ambivalence of the condemnation, might be read as a symptom of a worry that rhetoric might turn out to be a big problem if it is *not* so dispatched. In this paper I want to suggest that it is indeed a big problem for Locke, insofar as it is coterminous with the problem of language in general. This argument implies that rhetoric escapes from the frontiers imposed on it in section 34 of chapter X of Book III, and permeates the whole of Book III, insofar (and that is not very far) as that is where Locke's discussion of language is located. As it can and will be argued that on Locke's own terms language as a problem cannot be confined within the frontiers of Book III, but invades the *Essay* as a whole, then rhetoric will expand to become a crucial problem in Locke's whole enterprise. I shall argue that rhetoric is in fact an undermining and eventually deconstructive force at work in the most basic elements of Locke's account of knowledge. As is the way with rhetoric, this argument seduces me and arouses my desire; this desire is ambivalent with respect to Locke, but I take comfort from the fact that, in part at least, it is his desire too – for in the 'epistle to the Reader', he encourages me by saying: 'If thou findest any thing wanting, I shall be glad, that what I have writ, gives thee any Desire, that I should have gone farther'.[15] Whether the object of this desire is the 'Delight' and 'Pleasure' Locke ascribes to the understanding's 'searches after truth', and whether such pleasure is indeed to be located in great part in 'the very pursuit', are questions I leave open.

As I shall not have the time to support the possibly scandalous claim that rhetoric is basic to Locke's *Essay* as fully as it could be supported, I shall begin by specifying some things I think I shall *not* be doing, and by situating at least vaguely the position from which this text speaks.

First, I shall not be relying very much on Leibniz's comment in the *Nouveaux Essais sur l'Entendement Humain* to the effect that Locke, in denouncing rhetoric, does so eloquently, and thus combats eloquence with its own arms,[16] nor on an extension of that point which would argue that the whole of the *Essay* (and philosophy in general) necessarily relies on rhetorical strategies to get its points across. Nor, for the extension of the argument about the importance of language to the whole *Essay*, shall I simply use the banal point that language *must* be important to the *Essay* as a whole insofar as it is, necessarily, written in language. I do not think that any of these arguments is negligible, but simply that in their apparent self-evidence they offer little purchase on the problems at hand: the particular rhetoric of the denunciation of rhetoric could always be explained as a momentary lapse on Locke's part, or as a controlled example of playfulness, in neither case affecting his general theses concerning language and knowledge. As for the general rhetoric of philosophy in this type of argument, it could be hived off from the rhetoric of persuasion by the use of oppositions such as that of rhetoric and dialectic, persuasion and conviction, figure and concept, and so on. The possible deconstruction of these oppositions cannot be presupposed, and will depend on the location of a force of rhetoric prior to the terms they deploy.

Finally, although I do not want to deny that I approach Locke's text from the side of literature, as it were, rather than from the side of philosophy, and am therefore predisposed to see philosophy as something like a literary genre, the literal as a secondary effect of the figural, the concept as 'done for' by writing,[17] this 'literary' approach is not made on the basis of comments such as that of Nidditch in the Foreword to his edition of the *Essay*, to the effect that the book 'has long been recognized as one of the great works of English literature of the seventeenth century' (p. vii). My concern is not with the institutional 'recognition' of the book as 'really' literary or not, with the question of, for example, whether it should be studied in either philosophy or literature departments, or both, and to what extent, and so on, but with the existence, or, better, the *insistence* of something like 'the literary' or 'the rhetorical' in the most 'philosophical' or 'conceptual' elements of Locke's text. The 'something like' in this formulation should not necessarily be taken to be a wishy-washy 'literary' or 'rhetorical' approximation, a second-best to definition and rigour. It is, rather, dictated by the type of rigour it also questions, and points to the complex space between, the *entre-deux* of the philosophical and the literary, the literal and the figural, which is the concern of this work.

This positioning of the analysis is neither an invitation to benign interdisciplinary dialogue between literature and philosophy, nor a simple dismissal of philosophy in favour of literature, insofar as philosophers have been aware of the figural in Locke. M. R. Ayers, for example, refers to Locke's 'terminology, if that is not too grand a term for what sometimes seems almost an affectedly imprecise and figurative way of expressing himself'.[18] But neither is it a simple invitation to dialogue, insofar as this recognition is seen by philosophers as an unfortunate obstacle to thought and precision, something to be deplored, as does Peter Alexander in a reference to 'Locke's well-known lack of clarity of expression'.[19] The attempt of philosophers to get at 'what Locke had in mind', either to dismiss whatever that was through a generalised teleological violence in the name of the Truth,[20] or to reconstitute the system of thought revealed, once the ambiguities are 'ironed out by reference to the context', as Ayers has it, commits them to a notion of reading of the most simplistic kind, and, more importantly, to a view of language as an ideally transparent vehicle of meaning which can claim no real privilege of sophistication over Locke's own view of language. Locke's brooding over the fact that '[his] meaning . . . is often mistaken' in that part of the 'Epistle' added for the second edition,[21] and Ayers's comment that, 'Whatever Locke's *substratum* is, if he wrote *compos mentis,* it cannot be an entity that is undifferentiated, or 'other than' its properties, in fact',[22] share important presuppositions about how language works, failing to pose the question of what might be called, in Derridean fashion, the *necessary possibility* of meaning being 'mistaken', and the undecidability and ultimate irrelevance of whether Locke, or Ayers, or myself come to that, 'wrote *compos mentis*'.

A moment ago I suggested that the problem of rhetoric is 'coterminous with language in general'. I shall begin to approach that assertion through a consideration of general language. This is an attractive point of entry in that the doctrine of 'general ideas' and 'general words' has been recognised, at least since Berkeley's *Principles,* to be a difficult area of Locke's account of language, and also because the problem of general terms, and most acutely the 'names of mixed modes', on Locke's own terms, immediately inserts the problem of language into the heart of knowledge, and implies the necessity of a drift away from a notionally ideal propriety of terms on which any anti-rhetorical or anti-figural theory of language must depend. Just as Locke must ground his account of the origins of knowledge in the passive reception of discrete simple ideas by the mind, so he must attempt to ground his account of language in the dream of a language which would be a proper or literal nomenclature of those simple ideas. The attraction of this dream, and also its madness (and this madness at the very origin of Locke's theory already suggests that to wonder if Locke 'wrote *compos mentis*' might be to beg the questions both of writing and of madness), is well illustrated by Borges in his story, 'Funes the memorious':

> Locke, in the seventeenth century, postulated (and rejected) an impossible language in which each individual thing, each stone, each bird and each branch, would have its own name; Funes once projected an analogous language, but discarded it because it seemed too general to him, too ambiguous. In fact, Funes remembered not only every leaf of every tree of every wood, but also every one of the times he had perceived and imagined it. He decided to reduce each of his past days to some seventy thousand memories, which would then be defined by means of ciphers. He was dissuaded from this by two considerations: his awareness that the task was interminable, his awareness that it was useless. He thought that by the hour of his death he would not even have finished classifying all the memories of his childhood.[23]

Two vital, interrelated points are introduced here, beyond the immediate illustrative force of the passage: I shall have to be content to let them work all but silently on what follows. Both involve problems of temporality: the first asks questions about the time of a simple idea (and for example about Locke's definition of the 'Instant' as '*that which takes up the time of only one Idea* in our Minds'),[24] the second suggests on the one hand an originary *delay* of language in its definitions and classifications, and on the other an excessive *productivity* of language, which would never be able to catch up with the business of re-naming and classifying all the memories provoked by the first activity of naming and classification.

Even allowing Locke that his projection of a proper nomenclature is not *already* 'too general', it is clear that, as Locke recognises, such a language (admitting for the moment that it is 'proper' to call such a collection of names a 'language') is neither possible nor useful. The argument for its impossibility is not strictly philosophical, insofar as Locke ascribes it to a merely *de facto* limit in 'the power of Humane capacity'.[25] The account of the uselessness of such a 'language' is more interesting, and is subdivided into two parts. The first of these in fact provides a more convincing philosophical reason for the *impossibility* of this collection of proper names than did the argument designed to demonstrate that impossibility. Men, it is argued, use language (chiefly) to communicate their ideas. The use of a language of proper names applying to a person's (own, proper) ideas would not achieve the end of communication:

> This cannot be done by Names, applied to particular Things, whereof I alone having the *Ideas* in my mind, the Names of them would not be significant, or intelligible to another, who was not acquainted with all those very particular Things, which had fallen under my Notice.[26]

The second argument for the uselessness of such a view of language is that it would not help much in the way of knowledge. For although Locke grounds knowledge in simple ideas of sensation, he constantly links it to generality,[27] and thereby to abstraction, for, as is well known, it is through abstraction that ideas become general.[28] This process of generalisation is not, in the account of general ideas, given the type of economic motivation it is given in the discussion of general terms and general propositions, but is what 'properly' distinguishes men from animals: animals do not, according to Locke, have the capacity for abstraction and generalisation – and as he infers this from the fact that 'they have no use of Words, or any other general Signs' this suggests, as Berkeley was well aware,[29] that there is a more intimate connection between abstraction and language in general than the ostensible link of it to general language would allow for. (Incidentally, in a nicely ambiguous statement, Berkeley is able to suggest simultaneously that problems of generality are intimately bound up with language *and* that it is precisely language which has led Locke astray into his doctrine of abstraction and general *ideas*: 'if there had been no such thing as speech or *universal* signs there had never been any thought of abstraction'.[30] To show that Berkeley's and, through him, Hume's account of general terms does not solve the problems posed by Locke's position, but simply shifts into problems inherent in a different notion of representation, lies outside the scope of this paper.)[31]

That abstraction is in fact coeval with *any* use of language (and this is why Funes found Locke's notion of a nomenclature still 'too general') becomes clear in Locke's genetic account of general words. For the apparently (and only apparently) simple proper names from which he sets out (the example is of a child using the words 'Nurse' and 'Mamma' as proper names for individuals, and I leave aside the problems raised by the possible incompatibility of the two examples) already presuppose the abstraction implicit in the ability to link the several temporal perceptions of Mother and Nurse under the same name. The 'normal' sense of a proper name is already highly abstracted with respect to the notional infinite propriety of a proper name for every occurrence of an *idea*. If it could be shown that the use of such abstracted proper names is all that ties together the various occurrences of the 'same idea', then a fair number of Locke's presuppositions would be unsettled.

But it could be argued that even such a demonstration would not be very worrying for a doctrine of propriety insofar as no-one would take Borges' infinitely 'proper' language seriously (least of all Locke), and that the attenuated notion of propriety to a general idea in the mind (or, in Berkeley's revision, to a particular idea representative of a set of particulars) is perfectly defensible. All that would be needed in such a defence would be a certain directionality in the model of signification, posing, in the Lockean version, first things, then ideas of things, then words arbitrarily but consistently attached to those ideas (or, in the Berkeleyan version, first ideas, then words, or in the more 'commonsense' version, first things, then words). As long as the hierarchy implied in this type of listing is respected, the argument would go, then the notion of propriety, of the literal, is intelligible, and philosophy as a discourse can be protected against the aberrations of rhetoric. It is the desire to preserve the priorities of this type of model which can be read as determining Berkeley's shrinking back, at the end of the Introduction to the *Principles,* from the possibility of a generalised rhetoric which his attack on Locke's doctrine of abstraction had led him to envisage:[32] and he is right to shrink back, insofar as only these priorities can preserve philosophy from what can *provisionally* be called 'the play of the signifier' which would, in Locke's terms, make monkeys or parrots of us all and, more generally, ground the literal, and with it philosophy, no longer in reassuringly simple ideas, but in a generalised and originary non-propriety of which the use of language claimed by philosophy would be merely a local effect.

Now this type of reversal in Locke's model of language is precisely what happens in the case of what we should more normally think of as abstract terms, and specifically in the set of abstract terms which Locke calls 'names of mixed modes'. In Locke's terminology, 'modes' are 'such complex *Ideas,* which however compounded, contain not in them the supposition of subsisting by themselves', and are opposed to 'complex ideas of substances' which 'are such combinations of simple *Ideas,* as are taken to represent distinct particular things subsisting by themselves', and to 'relations', which are complex ideas and 'consist in the consideration and comparing one *Idea* with another'.[33] Within the category of modes, simple modes are made up of the 'same' idea repeated (for instance, the idea of 'two' is a simple mode, made by repeating the idea of a unity),[34] and mixed modes are made up of 'Combinations of simple *Ideas* of different kinds'.[35] Mixed modes and their names immediately problematise Locke's valorised, directional model of signification insofar as their 'originals' or referents are not located outside the mind, which could begin by receiving the ideas of them passively, but are made up by the mind actively. For it is at least possible (and Locke thinks it more usual) that the idea of a mixed mode such as 'hypocrisy' might first have been 'framed' without any 'pattern' of events or behaviour being prior to that framing. The problems this might provoke become clearer when Locke returns to the question of the *names* of mixed modes. By an apparent clumsiness of a type very common in the *Essay,* and not without its bearing on the questions at hand, Locke begins his discussions of this naming in Book III by repeating the discussion of the *Ideas* given in Book II; in the case of mixed modes at least, this repetition is not merely an *aide-mémoire,* a reminder of the doctrine of the idea to which the doctrine of the name could be subsequently attached, for here the name is not *superadded* to the idea, but is integral in the making of that idea. Locke has just repeated the point that mixed modes do not suppose some real existence of what they represent. He quickly reminds the reader that this does not imply that the ideas making up a mixed mode are not, as always, the simple units furnished by sensation and reflection, and goes on to describe how mixed modes are 'made':

> Wherein the Mind does these three things: First, It chuses a certain Number [of ideas previously in the mind]. Secondly, It gives them connexion, and makes them into one *Idea*. Thirdly, It ties them together by a Name.[36]

This is Locke's attempt to maintain the purity of his general model of signification. But the third stage of the process, the naming, is clearly vital to the whole:

> For the connection between the loose parts of those complex *Ideas,* being made by the Mind, this union, which has no particular foundation in Nature, would cease again, were there not something that did, as it were, hold it together, and keep the parts from scattering. Though therefore it be the Mind that makes the Collection, 'tis the Name which is, as it were the Knot, that ties them fast together.[37]

This does not merely imply, as Locke recognises, that 'now that Languages are made, and abound with words standing for such Combinations, *an usual way of getting these complex* Ideas, *is by the explication of those terms that stand for them,*[38] although this recognition[39] does relegate the previous genetic account to the status of an anthropological fiction: rather it introduces language into the very process of knowledge, for given Locke's account of the temporality of ideas, it is clear that the length of time the bundle of ideas could be held together by the Mind without the securing knot of language is unthinkably small. And insofar as the only reason Locke can give for the naming (and thus the creation) of mixed modes is communicative and thereby *social,*[40] then what is introduced into the ideation of the individual knowing subject by the mixed mode is the disruptive alterity of the social.

This introduction of the social into thought is fundamentally ambivalent. Insofar as the avowed priority of the name in this case seems to open up the use of mixed modes to the charge of psittacism, or monkeying about with words, on which Locke is elsewhere so prepared to pour scorn, then the temptation might be to minimise the importance of such terms and ideas. But given that questions of morality in general are questions concerning mixed modes, and given further that Locke declares that '*Morality* is *the proper Science, and Business of Mankind in general*',[41] then that temptation is evidently to be resisted. Further, in that Locke's discussion of mixed modes turns on examples necessarily involving *legislation,* it seems that the irruption into knowledge of language, the social and the moral necessarily involves a *scene of legislation* which is a heterogeneous element in an account of knowledge, and one not easily to be reduced. I wish to stress this last point as a means of taking a certain distance from Paul de Man's discussion of Locke in his essay 'The epistemology of metaphor',[42] which otherwise has a good deal in common with what I am attempting to elaborate here: where de Man typically makes an intrepid move to the literary, listing the examples Locke gives and commenting, 'The full list of examples . . . sounds more like a Greek tragedy than the enlightened moderation one tends to associate with the author of *On Government*',[43] I should want to stress the importance of the scene of legislation implied by the mixed mode, which is a scene of laying down the law as, of and in the divided place where language and thought and the social meet.

This importance of a scene of legislation is clearer still in Locke's account of the complex ideas he calls 'relations'. Even in 'natural relations' (for example, that holding between 'begetter' and 'begotten') Locke says that 'Mankind have fitted their Notions and Words to the use of common Life, and not to the truth and extent of things', and goes on:

> It is very convenient, that by distinct Names, these Relations should be observed, and marked out in Mankind, there being occasion, both in Laws, and other Communications one with another, to mention and take notice of Men, under these Relations: From whence also arise the Obligations of several Duties amongst Men.[44]

It is precisely these 'moral relations' that link mixed modes to Rule and Law and thus create 'Good' and 'Evil' out of the pleasure or pain we can expect to result from the reward or punishment which follows our observing or breaking the law.[45] Language is of vital importance in all this, as by naming actions in terms of mixed modes and relations we create 'the Mark of the value we set upon them'. The mixed mode is the 'positive idea' of the action, the relation is what refers it to a rule or a law: in the idea of 'stealing', for example, the relation is the element which determines it as crime or sin. In fact, this normative element called 'relation' must *already* be in the mixed mode to which Locke tries to suggest that it is added, insofar as the mixed mode only prevents itself from 'scattering' or dispersing because of the name 'stealing' which ties it together, and 'stealing' is already the name of a crime. This is no doubt why Locke returns for only two lines to the *names* of relations, saying only, 'What has been said here of mixed modes, is with very little difference applicable also to Relations'.[46]

Evidently these problems could be extended to Locke's political writings, insofar as the linguistic questions posed by mixed modes and relations imply some account of the social. This would be the place to examine, for example, the complicity between the priority of *propriety* in the *Essay* and that of *property* in the *Treatises,* the right to propriety of the names of simple ideas (a propriety which can be asserted but not defined) being homologous to the right to property (of his own body in the first intance) asserted of the pre-social liberal individual. And it would no doubt be possible to trace in Rousseau, for example, a similar relation of complicity between the claim that property is *not* naturally given and the claim that language is *not* originally proper but figured.[47]

There is a certain euphoria in Locke's account of mixed modes, a euphoria uniting the threats and promises generated by the new priority of language at this point, a priority separating mixed modes from the messy and disappointing concerns of complex ideas of substances, which retreat into the unknowable. The attraction of mixed modes is that as they are linguistic constructs with no 'pattern' in reality, they may be totally circumscribed and known: in other words, their real and nominal essences coincide. This allows Locke to claim, for example, that '*Morality is capable of Demonstration,* as well as Mathematics'.[48] Here the possibility of knowledge is in the primacy of the word.

Mixed modes, then, seem to open up the possibility of a valorised priority of language which disrupts to some extent the directionality of Locke's basic model of signification. The special case of the mixed mode would seem to incur the sort of denunciation Locke elsewhere makes of such a priority of the word:

Concerning Words also it is farther to be considered. *First,* That they being immediately the Signs of Mens *Ideas*; and, by that means, the Instruments whereby Men communicate their Conceptions, and express to one another those Thoughts and Imaginations, they have within their own Breasts, *there comes by constant use,* to be such *a Connexion between certain Sounds, and the* Ideas *they stand for,* that the Names heard, almost as readily excite certain *Ideas,* as if the Objects themselves, which are apt to produce them, did actually affect the Senses . . . *Secondly,* That though the proper and immediate Signification of Words, are *Ideas* in the Mind of the Speaker; yet because by familiar use from our Cradles, we come to learn certain articulate Sounds very perfectly, and have them readily on our Tongues, and always at hand in our Memories; but yet are not always careful to examine, or settle their Significations perfectly, it *often* happens that *Men,* even when they would apply themselves to an attractive Consideration, do *set their Thoughts more on Words than Things.* Nay, because Words are many of them learn'd, before the *Ideas* are known for which they stand: Therefore some, not only Children, but Men, speak several Words, no otherwise than Parrots do, only because they have learn'd them, and have been accustomed to those sounds. But so far is there a constant connexion between the Sound and the *Idea*; and a Designation, that the one stand for the other: without which Application of them, they are nothing but so much insignificant Noise.[49]

What happens then is this: in this type of psittacism, as in the 'proper' use of mixed modes, language, which should be a secondary adjunct to ideas, invented simply for the communication of those ideas which could exist in their purity in a pre-social and pre-communicative isolation, begins to come first. It is now the word which precedes the idea and dominates thinking – the transparency to itself of personal meditation among ideas becomes contaminated by its means of communication to others. The separation from the other which made communication necessary (always already necessary, insofar as God 'designed Man for a sociable Creature', and 'furnished him also with Language, which was to be the great Instrument, and common Tye of Society')[50] here becomes a separation of the oneself from the other-in-oneself: the thinking self becomes primarily a 'listening' self, 'alienated' from self-possession of thought because of this 'listening' to the 'sounds' of language. Locke's later mention of a second use of words for recording one's own thoughts as a help to one's memory only stresses this disruption of the other in the self. It remains to be shown that this 'other-consciousness' which enters into the proper conception and use of the mixed mode and haunts all language as the possibility of its abuse (though this particular abuse does not fit easily into Locke's list, but is rather the condition of possibility of all other abuses), inhabits, from the start, self-consciousness and thereby, for the two go together in Locke, consciousness itself.

Let us return first from the problematic heights of mixed modes, retaining the suspicion they have introduced that language is bound up with the possibility of knowledge in a more intimate way than Locke would like, to the atomic units of those mixed modes, which are still, of course, simple ideas of sensation and reflection. The attempt to maintain propriety, and the secondarity of language which propriety implies, must ultimately depend on arguments conducted at this level. Locke's account of why the aggregate of proper names for simple ideas is impossible and useless omits a crucial reason for this impossibility and uselessness. For if there *were* such a language, it would have to be a language without propositions at all. If every proper name and every idea were really discrete, isolated in its propriety or enclosed in its property, no other name could be linked with it. The very attempt to make such a link would presuppose the addition to the aggregate of proper names of at least one more item, namely the copula, and such an addition already breaks open propriety, immediately by introducing this heterogeneous particle into the aggregate, and secondly by presupposing the possibility of relations of identity and/or difference between the elements of that aggregate. Even if the role of the copula were purely negative, used only in propositions of the type 'A is not B', 'B is not C' and so on, then a minimal differential articulation is posited, and this is in fact the only way in which the supposed propriety of the terms could conceivably be maintained in the first place. And if the copula is used only to assert the self-identity of terms, in the form 'A is A' (and Locke recognises that there are as many self-evident propositions of this type as there are simple ideas), then this is, on Locke's account, only a 'trifling' proposition, presupposed by the prior recognition of the idea: and it also, we might add, presupposes the type of temporal alterity suggested by Borges' fable, and it is only this alterity which allows for the repetition of the idea as the 'same' idea and the name as its name. More importantly, this recognition of the presupposition of the possibility of the copula among the names of simple ideas, although it is itself not a name (and given by Locke only the briefest of discussions in that briefest of chapters called 'On Particles'),[51] it already disrupts the ideal propriety of terms in the same movement with which it makes possible anything like an affirmation of that propriety.

These observations suggesting that Locke's positioning of language with respect to ideation cannot be simple and unproblematical can now be extended to the account of the acquisition of the basic units of all knowledge through sensation and reflection. It is clearly vital to Locke's account that sensation be the prior term, insofar as the mind must be provided with at least one idea of sensation before it has anything to reflect about. Reflection is defined as 'the *Perception of the Operations of our own Minds* within us, as it is employ'd about the *Ideas* it has got'.[52] Accordingly, the first idea of reflection is the idea of perception,[53] there having first been perception of an idea of sensation.

Now it can be argued, and has been argued by Douglas Greenlee,[54] on whom I draw for this preliminary point, that insofar as for Locke ideation entails consciousness of ideation, then any idea must be accompanied by the idea of 'idea': as Greenless puts it, 'hence sensation, *from the beginning,* it follows, would have to be accompanied by the minimal reflective operation of the perception of having the sensory idea' – and he goes on to suggest that this means that the reflective idea of 'idea' is '*as old as* the idea of sensation'.[55] The formulations I have emphasised, 'from the beginning', and 'as old as' point nicely to the fact that there is no simple origin in Locke's account of origin. It is possible to radicalise this, and to argue that only the reflective idea (which is itself originally split insofar as it is also an idea of reflection) allows the sensory idea to exist as idea of sensation, and thereby gain any effect of propriety or identity. Clearly this implies a certain *marking* of the idea, a marking of it as different from other ideas, and this differential marking is a type of proto-language which is a condition of possibility for the proto-language of proper names of simple ideas, and can be assimilated to what Derrida calls '*archi-écriture*'.[56] This is another name for the split and never-present origin of ideas in Locke's theory, and suggests why it is unthinkable that any single or self-present idea could ever be present, or presented, to whatever august personage presides in what Locke preciously calls 'the mind's Presence-room'.[57] This split origin evidently disallows any simple notion of a *tabula rasa* or of a simple passivity of the mind. The mind is originally split *by* reflection, which becomes a condition of sensation. This split introduces something like 'activity' into the supposed first 'passivity' of the mind, and only this allows for the possibility of language, the use of which is thus grounded in a sort of primary 'abuse', insofar as it can never 'properly' name the non-proper *différance* which precedes and inhabits it. The split also suggests that the mind is always inhabited by alterity insofar as there are two agencies or instances at its beginning. Locke's secondary story about the usefulness of language in the self as a sort of *aide-mémoire,* which it will be remembered introduced the other into the self, can be read as a deferred figure of this originary alterity. There is always an other in the mind's presence-room.

This originary spacing is the space which allows for the necessary possibility of Locke's being 'misunderstood' (and for complaints that I have wilfully misunderstood him here, for example). It is also the space in which philosophy as a metalanguage tirelessly denouncing the 'abuse' of language called rhetoric, and rhetoric itself, are located. Locke depends on this originary drift from propriety for the possibility of doing what he does, which is thus a local activity within a generalised figurality. Locke needs the perfect cheat as the (feminine) matrix of his (virile) claims. Where Locke would like there to be a possible disjunction of the literal *or* figural, separated into their 'proper' places, there is only the literal *and* the figural, fighting over their common ground in a general rhetoric. There is no point in looking for agreement here, for the difference is original, and any supposed accord between the literal and the figural, and to that extent between the philosophical and the literary, is merely a secondary violence worked on this difference.

In the parallel reading of the political writing proposed a moment ago, the need for government as a supplementary instance to preserve (naturally achieved) property[58] would depend on a similar originary shift, which Locke can only read as a sign that man is degenerate.[59] The necessary possibility of violent misappropriation of others' property in Locke's otherwise benign state of nature dictates the formation of society and government as an instrument of reappropriation, just as the necessary possibility of the misappropriations of rhetoric dictates philosophy's legislative reappropriations of words to ideas. A point of entry into the common root of these problems might be provided by §143 of the *First Treatise,* where Locke quotes Filmer (whom John Dunn describes as having a 'strange, punning theory')[60] as saying: '*In the dispersion of* Babel, *we must certainly find the Establishment of Royal Power, throughout the Kingdom of the World*', and, after a few general comments, continues,

If our A. has no better Foundation for his Monarchy than a supposition of what was done at the dispersion of *Babel,* the Monarchy he erects thereon, whose top is to reach to Heaven to unite Mankind, will serve only to divide and scatter them as that Tower did; and instead of establishing civil government and order in the World will produce nothing but confusion.[61]

If, then, we are among those who, in Locke's words, 'pretend *seriously* to *search after,* or maintain *Truth*',[62] or at least pretend to pretend seriously, then we might keep up the pretence rather than the pretension. This is still quite a *serious* pretence, and one of the things it pretends seriously is that the pretension is not serious, *n'est pas sérieux*. The divided unity of this serious pretence and frivolous pretension, of the search which is also a maintenance, of the object of desire called 'Truth' which dissolves into, among other figures, both mother *and* whore, is perhaps namable as a '*perfect cheat*'.

Notes

1 All references to Locke's *Essay* are to the edition by Peter H. Nidditch, Oxford, 1975. The numbers given refer to Book, Chapter, and Section respectively. References to the Epistle to the Reader are to the page number.

2 Locke, *Essay,* II, X, 1.

3 *Ibid.*, III, X, 1.

4 *Ibid.*, III, XI, 3.

5 *Ibid.*, III, XI, 3.

6 *Ibid.*, III, X, 34.

7 *Ibid.*, III, XI, 1.

8 *Ibid.*, III, XI, 3.

9 *Ibid.*, III, X, 34.

10 *Ibid.*, III, X, 34.

11 *Ibid.*, III, X, 5.

12 *Ibid.*, III, XI, 4.

13 *Ibid.*, III, X, 34.

14 See the comments in *Ibid.*, II, XXXIII, 19.

15 *Ibid.*, p. 7.

16 G. F. Leibniz, *Nouveaux Essais sur l'Entendement Humain,* ed. Jacques Brunschwig, Paris, 1966, p. 306.

17 Cf. Derrida's untranslatable comment somewhere in *Glas,* Paris, 1974: 'Dès qu'il est saisi par l'écriture, le concept est cuit'.

18 M. R. Ayers, 'The ideas of power and substance in Locke's philosophy', in *Locke on Human Understanding: Selected Essays,* ed. I. C. Tipton, Oxford, 1977, p. 85.

19 P. Alexander, 'Boyle and Locke on primary and secondary qualities', in Tipton, ed. *Selected Essays,* p. 85.

20 I am thinking especially of Jonathan Bennett's *Locke, Berkeley,*

Hume: Central Themes, Oxford, 1971.

21 Locke, *Essay,* p. 11.

22 Ayers, 'The ideas', p. 79.

23 Jorge Luis Borges, 'Funes el memoriose', in his *Ficciones,* Buenos Aires, 1956, p. 125. English translation by James E. Irby in his *Labyrinths,* Harmondsworth, 1970, p. 93.

24 Locke, *Essay,* III, XIV, 10.

25 *Ibid.,* III, III, 2.

26 *Ibid.,* III, III, 3.

27 See for example, *Ibid.,* IV, V, 10. and IV, VI, 2.

28 *Ibid.,* II, XI, 9.

29 G. Berkeley, *The Principles of Human Knowledge,* in *Principles, Dialogues and Correspondance,* ed. Colin Murray Turbayne, New York, 1965, Introduction, §11.

30 *Ibid.,* §18.

31 The argument is developed by E. Husserl in §29 of the second of his *Logical Investigations,* tr. J. N. Findlay, New York, 1964.

32 Berkeley, *Principles,* Introduction, §20.

33 Locke, *Essay,* II, XII, 4, 6, 7.

34 *Ibid.,* II, XIII, 9.

35 *Ibid.,* II, XXII, 9.

36 *Ibid.,* III, V, 4.

37 *Ibid.,* III, V, 10.

38 *Ibid.,* II, XXII, 3.

39 Repeated in *Ibid.,* III, V, 15.

40 *Ibid.,* III, V, 7.

41 *Ibid.,* IV, XII, 11.

42 Paul de Man, 'The epistemology of metaphor', *Critical Inquiry,* 5, 1978–9.

43 *Ibid.,* p. 21.

44 Locke, *Essay,* II, XXVII, 2.

45 *Ibid.,* II, XXVIII, 5.

46 *Ibid.,* III, V, 16.

47 The first of these claims is made especially in the *Discours sur l'Origine de l'Inégalité* and the second in the *Essai sur l'Origine des langues.*

48 Locke, *Essay,* III, XI, 16.

49 *Ibid.,* III, II, 6–7. See in addition III, XI, 24.

50 *Ibid.,* III, I, 1.

51 *Ibid.,* III, IV.

52 *Ibid.,* II, I, 4.

53 *Ibid.,* II, IX, 1.

54 Douglas Greenlees, 'Locke's idea of idea', in Tipton, ed. *Selected Essays.*

55 *Ibid.,* p. 44. My emphasis.

56 See for example *De la Grammatologie,* Paris, 1967, and *Positions,* Paris, 1972.

57 Locke, *Essay,* II, III, 1.

58 J. Locke, 'Second treatise on government' in *Two Treatises on Government,* London, 1884, §94 and §124.

59 *Ibid.,* §128.

60 John Dunn, *The Political Thought of John Locke,* Cambridge, 1969, p. 75.

61 Locke, 'First treatise . . .' in *Two Treatises,* §143.

62 Locke, *Essay,* III, XI, 3.

7 *Geoffrey N. Cantor*

Weighing light: the role of metaphor in eighteenth-century optical discourse

7.1 *Introduction*

According to conventional wisdom the language of science is a transparent medium which, unlike poetry, political oratory or music, functions solely as a conveyer of information. It is this conception of scientific language which has led most commentators, especially philosophers of science, to equate scientific with propositional discourse. Moreover a stark antithesis has generally been drawn between this form of discourse and rhetoric, which is considered as having no legitimate role in science. This antithesis, between the propositional or literal and the rhetorical, found its classic formulation in Locke's *Essay Concerning Human Understanding* where he claimed that rhetoric aims 'to insinuate wrong *ideas*, move the passions, and thereby mislead the judgment, and so indeed are perfect cheat'.[1] Rhetoric, Locke continued, is to be avoided 'in all discourses that pretend to inform or instruct . . . and, where truth and knowledge are concerned, cannot but be thought a great fault either of the language or person who makes use of them'. A few lines later he dismissed rhetoric as 'that powerful instrument of error and deceit'. Built into this passage is the contrast between rhetoric and another type of discourse which informs, instructs and conveys truth, but does not mislead the judgment, inflame the passions or propagate error. In this other discourse words signify ideas, thus by using the medium of words a speaker or author can convey the ideas in his/her mind to the mind of the listener or reader. Such a process facilitates effective social intercourse, but Locke particularly emphasised that this process of replication played an essential role in science where all forms of obscurity and error had to be eliminated. Indeed for Locke, as for Sprat and many other commentators, one of the essential features of science as rational discourse is that it is not rhetorical.

It is beyond the scope of this paper to trace the history of this great divide through to the present day. Suffice it to say that most scientists and philosophers of science during the last three centuries have accepted that science deals in propositions whose terms are concisely defined whereas rhetoric is proscribed.

However, contrary to this conventional wisdom one particular trope does seem to play some role in science. Metaphor, which has long been on the agenda of the literary critic, has during the last couple of decades become a contentious subject among philosophers, philosophers of science, aestheticians and psychologists. As an indicator of the revival of interest in metaphor several collections of essays have recently been published giving prominence to new and exciting research problems. As the editor of one such collected volume has written: 'We are in the midst of a metaphormania'.[2] Although this judgment may be true in other spheres, the history and philosophy of science has not been inundated with studies of metaphor; nevertheless, there is a small and growing literature on metaphor in science. On the historical side one can point to such studies as Robert Young on Darwin, Canguillem on physiology or Lelland Rather on medical metaphors as examples of this genre.[3] Less easy to classify are eclectic studies such as Stephen Pepper's *World Hypotheses* or Roger Jones's recent *Physics as Metaphor* both of which, while being exciting and suggestive, employ metaphor in a loose and ahistorical way.[4] More disciplined and analytical are the writings of philosophers of science such as Mary Hesse, Rom Harré, Richard Boyd and T. S. Kuhn. Among these philosophers of science there is a consensus that when it comes to science the theory of metaphor which is of most use is some form of interactionist theory. This theory dates from I. A. Richards's *Philosophy of Rhetoric* of 1936 and was subsequently taken up by Max Black in the mid-1950s.[5]

The proponents of the interactionist theory are concerned, *inter alia,* with the problem of how metaphor works. Two constituent elements are posited – what Richards called the 'tenor' and 'vehicle' which others have denominated respectively the primary or principal subject and the secondary subject. For an effective metaphor the two subjects need to be taken from conventionally different systems, domains or modes of discourse, but are brought together interactively in the metaphor. When, for example, Boyle wrote that nature is like the grand clock at Strasburg[6] he was bringing together concepts from two conventionally-different domains – the God-created world of nature, and the product of human artifice. Metaphoricity does not depend on the phonic or syntactic structure of the sentence; instead, terms are conjoined which don't normally hunt together but are derived from different types of discourse and exhibit both similarities and differences. Thus with the Boyle example the sweep of the clock's hands mirror the movements of the sun, moon and planets. However, unlike nature, the clock requires regular winding and mending.

According to the interactionist theory metaphor involves a tension between the two subjects or, probably more correctly, a 'strain' between the speech act and the total linguistic environment. There is a shift in the meaning of the primary subject as some of its qualities are accentuated and others downplayed. The clock metaphor draws attention to those of its parts, like the astronomical movements, which are clockwork-like in their regularities. It also urges one to find the 'springs' and 'cogs' producing these regularities and to interpret nature through mechanical principles.

One other point needs to be clarified at this stage: scientific propositions cannot be formally distinguished from metaphors. A sentence such as, 'Light is a volley of small particles of matter' is a proposition, and in adopting a propositional reading we are led to ask whether the proposition is true or false, and how we might test it. On the other hand, the same sentence can function as a metaphor. At stake now are not so much questions of truth, verification or confirmation, but the meanings of the primary and secondary subjects when brought into an interactive relationship. As has often been stressed, metaphoricity cannot be ascertained by considering the sentence *in vacuo* but instead we must examine the key terms in their contemporary cultural context in order to decide whether or not they interact through tension. To identify metaphor, then, we must ask an essentially *historical* question. We must ask what meanings and implications are associated with both the primary and secondary subjects and the effect of bringing them together in a specific context. At other times and places the same set of words may produce a very different metaphor or even be devoid of metaphorical content; indeed, we usually find that metaphors derived from an alien culture are opaque to us. Since metaphor is context-dependent, the task of analysing metaphors clearly falls to the historian.

Among those who have sought to show that scientific language is or trades on metaphor there seem to be two recurrent strategies. One such strategy derives from those radical theorists of language who challenge the traditional dichotomy between the literal and the figural. For Nietzsche and Derrida, for example, all language is essentially metaphorical and therefore the language of science, so often considered a special case, can no longer be pronounced metaphor-free. The radical argument is both highly interesting and provocative, and although I shall return to it briefly at the end of this paper I shall concentrate on the alternative strategy, which has been articulated by philosophers concerned with providing theories of meaning and reference.

According to these philosophers metaphor can perform three functions in science.[7] The first of these functions is called by Boyd the generative function. Implied here is something much stronger than the claim that metaphor plays a role in the context of discovery – to use Popper's term – and then subsequently disappears from science. Instead, Boyd speaks of *theory constitutive metaphors* and argues that any worthwhile scientific theory is continually being extended through articulation of its underlying metaphor or metaphors.[8] Metaphor thus plays a creative role in directing research into new areas through the discovery and articulation of both the primary and secondary subject domains. Thus theory-constitutive metaphors are not mere ornaments and are certainly not the source of deceit, but are essential to the development of scientific theory.

A second role attributed to metaphor is in the framing of scientific terminology. This topic has recently attracted the concern of realist philosophers of science who, dissatisfied with the definitional account of reference fixing, have utilised metaphor centrally in their discussions of non-definitional reference. As with the generative function, the reference function of metaphor operates not only in new or immature theories but also in mature theories as they are articulated through research. Thus the two foregoing functions of metaphor bring metaphor not only into Hannaway's pre-Lavoisian chemistry, but also into the research laboratory.[9]

The third and final role of metaphor is not concerned with research but with teaching.[10] In this didactic context new ideas and theories are introduced in terms of concepts which are familiar; for example, the Bohr atom is often taught by appealing to the more widely-known model of the solar system. Frequently this didactic role for metaphor is portrayed as an unfortunate but temporary expedient which can soon be dispensed with once students obtain a firm grasp on their subject. This view seems, however, less than satisfactory since these didactic metaphors possess a permanency in scientific discourse through new cohorts of students being initiated into the science. Moreover, this view sets up an artificial distinction between learning and research activities.

The above considerations suggest that far from being inimical to science, metaphor plays a central role and one that the historian of science should analyse. In the remainder of this paper the role of metaphor will be examined in a number of eighteenth-century texts dealing with the nature of light. Since 'light' inhabits not only natural philosophy texts but also other forms of eighteenth-century discourse, most notably theological discourse, we can expect light metaphors to carry resonances from outside science. One of my major concerns will be to demonstrate the theological origin of many eighteenth-century optical metaphors and thus the theological underpinnings of optical discourse. My other main aim is to analyse the several different functions of light metaphors in this specific historical context.

7.2 *Metaphor and the fluid theory of light*

According to a sizable group of eighteenth-century theorists, light is a fluid flowing from the sun towards the stars, this flux of light illuminating the earth and other celestial bodies. Usually this fluid was conceived as particulate, although the trajectories of individual particles were not discussed. Instead, fluid theorists concentrated on the combined progressive motion of these particles which behaved like water flowing in a river. At the heart of this theory was the proposition 'light is a fluid,' and yet if we read these four words solely in the propositional sense we foreclose further discussion of their contextual significance. There is an inherent ambiguity in reading the text; are we dealing with a proposition or a metaphor? Perhaps some mixture of the two? Were these theorists appealing to fluids in order to illustrate to the reader their theory of light? Or perhaps they tried to legitimate a metaphorical understanding of the world by dressing their metaphor in propositional clothing? By raising these questions I hope to have shown the indeterminacy of the above sentence. Only by examining the contexts in which it occurred and the meaning attached by specific authors to the word 'fluid' can we decide between the above alternatives. In the following analysis we will see that 'fluid' connected optical discourse, not only with hydrokinetics, but also with revealed theology.

Throughout their writings fluid theorists articulated metaphors which drew light into a form of discourse which dealt with the motion of a fluid, especially water. 'Some philosophers', wrote William Jones, 'have contended for an actual progressive motion of the parts [of light], as in the stream of rivers.' John Lacy referred to the matter of light as being 'in a fluid Form' while Peter Leslie called light a 'subtile, active fluid' and Charles Wilkinson pronounced light to be a 'fluid sui generis'. Another late eighteenth-century natural philosopher Felix O'Gallagher spoke of 'this body of light, though ever flowing from the sun' and the 'flood of solar matter'.[11]

At one level it might be argued that such metaphors perform the function of allowing one to talk about the behaviour of light in terms taken from a more commonplace set of phenomena. While this is certainly true the above thesis ignores the crucially important role of theological discourse which, I want to argue, provided a linguistic structure in which both light and fluids could be articulated. Before turning to these issues it should be noted that the fluid theorists, unlike the supporters of the rival theories, were strong advocates of *revealed* theology.[12] Also it is important to appreciate that light is to be found both in the book of nature and in the book of revelation such that the word 'light' inhabits the discourse of both physics and theology. Thus in the Bible there are several light metaphors which portray God – or Jesus in the case of the New Testament – as the source of spiritual light which can illuminate the soul of the believer. 'The Lord is my light and my salvation' (Psalm 27:1); 'The Lord shall be to thee an everlasting light' (Isaiah 66:19); Jesus speaks of himself as 'the light of the world, he that followeth me . . . shall have the light of life' (John 8:12).

The next step in the argument takes us to the book of Jeremiah and to one of the most frequently-cited metaphors in the Bible. In two passages (Jeremiah 2:13, 17:13) God is referred to as the *fountain of living waters*. The spiritual light of life emanating from God is thus expressed by the flow of water and it is particularly important for our discussion that water can be attributed with vital, active, spiritual significance. We are thus faced with Biblical underpinnings for the intimate connection between light and water, since both express the spiritual aura emanating from God. The above passages from Jeremiah also indicate a further aspect of my argument since God is called the *fountain* from which the 'living waters' emanate. Furthermore, in the Trinitarian tradition God the Father has been represented by the sun and hence the sun, too, has assumed the office of a fountain. Psalm 84 contains the line: 'For the Lord God is a sun and shield: the Lord will give grace and glory . . .' Just as God spreads light (spiritual) so the sun spreads light (physical) throughout the universe. The word glory is another term which straddles physical and theological discourse. Chapter four of the prophetic book of Malachi performs a similar role. Here it is said that the 'sun of righteousness' shall arise in those who fear God; but this passage was frequently given a Christian gloss such that Jesus is called the 'sun of righteousness' who enlightens, quickens and comforts his people.

If one takes these metaphors seriously, as indeed they were taken in the eighteenth century, then the sun is no longer a physical object but the counterpart of one of the divine persons – usually God the Father – in the physical world. This insistence on fusing the physical and spiritual is to be found frequently in natural philosophy texts propounding the fluid theory. In discussing optics in his *Lectures on Natural and Experimental Philosophy* of 1794 George Adams reminded his readers that, 'Our LORD . . . is represented in the sacred oracles *as* the SUN, the fountain of LIGHT . . . [and] truth [is expressed] by the significant emblem of LIGHT, and the LIGHT OF LIFE'. Earlier in the century Duncan Forbes, the Scottish judge, enamoured with John Hutchinson's Biblically-orientated natural philosophy wrote that the 'Power of Irradiation in a spiritual Sense, as we see performed by the Sun in a material Way, is ascribed to the invisible God'. Another of Hutchinson's early followers, George Horne, spelled out the relationship between the effable and ineffable light thus: 'Light is the great source of blessing in the natural world, *love* in the moral world'.[13] Adams, Forbes and Horne were just three of the many eighteenth-century fluid theorists who conceived the sun – the fountain of light – as representing God the Father in the physical world and adopted an intimate relationship between the physical and spiritual (or moral) light.

The above evidence strongly supports the thesis that for the fluid theorists optical discourse was impregnated by metaphors derived from theology. However, this line of argument appears to encounter the following objection. With few exceptions, fluid theorists insisted that light is material, composed of a plenal fluid in motion; indeed they often argued that it was heretical to attribute non-material qualities – such as force – to matter. This objection exposed a central paradox in the fluid theorists' natural philosophy and in their mode of discourse. On the one hand, they sought a material world different from the non-material; to keep God ontologically separate from his creation; to make the physical world a grand machine which depended on the laws of mechanism and not those of force. On the other hand, like Croll's view nearly two centuries earlier,[14] the physical world was an image, a symbol of the non-material. While rationalist considerations underpinned the first of these considerations, language underpinned the second. It was by the word that God had created the universe. It was through language too that the divine plan was revealed, a plan in which the physical and moral bore a complete correspondence. Both types of light flowed similarly from the source, the difference being that one was material and the other immaterial. The metaphors of flow and the fountain were not fanciful analogies but were tropes which revealed the underlying similarities between the two forms of light. In the discourse of the fluid theorists God and matter, revealed theology and natural philosophy, co-existed uneasily.

7.3 *Metaphor in projectile optics*

Turning to projectile optics we encounter several different metaphors which were used extensively. The first of these – what might be called the root metaphor – related light directly to particles of matter.

7.3.1. *Root metaphor*

Materialist theories of light have a long history but it is to Newton's writings which we now turn to examine the metaphoricity of the projectile theory. In Book I of the *Principia*[15] Newton discussed the trajectory of a small material body through a parallel-sided region under the influence of an attractive force of constant magnitude acting perpendicularly to the side through which the body had entered. It is important to notice that Newton imported into this dynamical argument terms derived from optics; when certain conditions pertain the body is described as being reflected, and he also discussed its trajectory in terms of the angle of incidence and the angle of reflection. In the ensuing scholium Newton confronted the question of the relation between this dynamical problem and the strikingly similar optical phenomena of reflection and refraction. While noting that a strong analogy existed between the two cases, Newton asserted that he was not prepared to enter into the more fundamental question of whether rays of light were indeed material bodies.

The 'queries' to the *Opticks* constitute one of the most interesting scientific texts from a stylistic viewpoint. Using the interrogative form Newton could engage in speculation in public, often employing weak justifications; whereas elsewhere in the *Opticks* and in the *Principia* he dealt in demonstrations or strong inductive arguments. Also, the form of presentation made him less vulnerable to direct attack by his adversaries. In the *Optice of* 1706 he asked: 'Are not light rays very small bodies emitted from shining surfaces?'[16] The question is of course rhetorical and the affirmative answer is given further force by the several inductive arguments Newton offered to support the projectile theory and his outright rejection of the wave hypothesis.

We have to wait just four years for the linguistic transformation to occur. With the second volume of John Harris's *Lexicon Technicum* of 1710 Newton's query had become a positive statement about the nature of light: 'The *Rays of Light* are therefore certainly *little Particles* actually emitted from the *Lucent Body*'.[17] This transformation involves not just metaphor, but identity. Light had become matter. The discourse of matter had consumed optical discourse. The words 'therefore', 'certainly' and 'actually' function as locks to prevent light from escaping from the domain of matter. No such escape was possible, for works on natural philosophy soon reiterated Harris's proposition. If Newton's language left matters open, his followers did not. '*Light* is a Body'; it 'must be small Corpuscles'; 'light consists of very minute particles of matter'; 'light consists of an inconceivable great number of particles . . .'[18] The chorus is deafening. Rarely showing any failure of will, eighteenth-century writers consigned light to the realm of matter.

Just how metaphorical was the claim that light is projected particles of matter? According to the interactionist theory, the effectiveness of a metaphor depends on the tension developed between its primary and secondary subjects. As stressed earlier, we can only determine whether such a tension existed by examining the use of words in specific historical situations. The situation is further complicated in the present case since it is clear that the purported materiality of light was considered metaphorical by some writers and not others. While deferring discussion of the generative function of metaphor until Section 7.3.4, I will briefly characterise two opposing responses.

Many of the public lecturers stated that light is composed of small particles travelling at a high speed. The rhetoric of the lecture theatre left little room for equivocation and the lecturer simply stated what light is or consists of. This mode of presentation was thus propositional and not metaphorical; or, to state the situation in a different way, the materialist metaphor was a dead metaphor. However we should not necessarily assume that the metaphor was dead.

For at least one identifiable group there was a strong antithesis between light and matter (as generally conceived); these were people, often of deep religious conviction, for whom light was essentially spiritual. Three examples will suffice. While accepting that particles of matter are responsible for many physical phenomena, Christopher Smart insisted that colours – and thus light – were spiritual, not material. A second example is provided by George Berkeley who argued in *Siris* (1744) that light is a modification of that subtle medium – the soul of the world – on which all activity depends. Finally, in a short poem William Blake transformed Nieuwentijdt's sand metaphor (which is discussed below):

> Mock on, mock on, Voltaire, Rousseau,
> Mock on, mock on; 'tis all in vain!
> You throw the sand against the wind,
> And the wind blows it back again.
>
> And every sand becomes a gem
> Reflected in the beams divine;
> Blown back they blind the mocking eye
> But still in Israel's paths they shine.
>
> The Atoms of Democritus,
> And Newton's Particles of Light
> Are sands upon the Red Sea Shore,
> Where Israel's tents do shine so bright.[19]

For Blake the spirituality of light – the beams divine – displays the crass catachresis involved in 'Newton's Particles of Light'.

7.3.2 *The sand metaphor*

According to the projectile theory the particles of light must be exceedingly small, otherwise they would collide with one another and produce both chaos and confused vision. Moreover, it was clear that the number of particles emitted from even a small candle is very large. We do not have experience of these teeming light particles, but the small, profuse objects with which we are most familiar are particles of sand. Thus sand provided a suitable metaphor to convey the profusion of light particles. The immediate source of this metaphor is to be found in an early-eighteenth-century work on natural theology. Bernard Nieuwentijdt's *The Religious Philosopher* was translated by Desaguliers and appeared in English in 1718–19. The work's aim was to convince the atheist of his errors by showing him the structure of God's creation and arguing that it must have been framed by divine providence. When Nieuwentijdt discussed light, one of his immediate aims was to 'calculate' the number of light particles emitted from a candle. After adopting several assumptions and churning through the mathematics he arrived at the conclusion that $418{,}660 \times 10^{39}$ particles were emitted each second from a size six candle. 'From which it may be calculated that *in one Second* (which is commonly equal to one Pulse of an healthy Man) *there fly out of a burning Candle of Six to the Pound, many more Particles of Light than a thousand times a thousand Millions of that Sand the Number the Earth can contain, or be equal to*.' In order to warn the reader not to take this calculation literally Nieuwentijdt added, 'if we had kept to a strict [sic] Calculation, the Number thereof would very far, yea unconceivably, surpass what we have here set down'.[20] Thus far from offering a strict, accurate determination Nieuwentijdt's argument, including his use of mathematics, was rhetorical. His aim was to evoke wonder and surprise in the reader by reciting the superabundance of light particles emitted each second from a candle. Yet such wonder and surprise served another purpose, which was to show divine providence at work in every part of nature and thus to use science to combat atheism. As Desaguliers boldly stated in his introduction, a philosopher cannot be an atheist.

Even in Nieuwentijdt's book we see some of the tension created by the sand metaphor. Given the prodigious number of particles emitted from a candle each second, many more orders of magnitude must be involved in the emission of light particles from the sun each second, and several more orders each year. There loomed the danger of rapidly exhausting the sun. Two counter-strategies occurred to Nieuwentijdt. Either the matter emitted from the sun circulated and thus returned to its source, or 'the Particles of Light . . . [might] be conceived to be so very small . . . yet being taken all together, they may not perhaps amount to the Quantity of one Grain of Sand; for which Reason there can be no Diminution perceived in the Body of the Sun in as many more Years or Ages'.[21] Whereas in the earlier discussion Nieuwentijdt used grains of sand as a figure of speech to stand for superabundance, now he in effect compressed those inconceivably numerous grains into a single grain of sand. The 'one Grain of Sand' became a figure of speech for the inconceivably small bulk (and thus mass) of the light particles.

Despite the problems manifest in Nieuwentijdt's pseudo-calculation, his figures and the sand metaphor were frequently employed by other eighteenth-century natural philosophers. Thus, on being asked to discuss the smallness of the light particles, the tutor in Benjamin Martin's *Philosophical Grammar* replied: 'It is computed that in a Second of a Minute, there flies out of a burning Candle, the following Number of Particles of *Light*, 41866000000000000 00000 00000000000000000000000, which is 10 000 000 000, or ten millions of millions Times a bigger Number than 100000000000000000000000000000, the Number of Grains of Sand, computed to be contained in the whole Earth'.[22] Likewise several other texts, such as James Ferguson's widely-read *Lectures on Select Subjects*, cited without comment Nieuwentijdt's *calculation* of the number of light particles.[23]

This example illustrates two interesting features. Firstly, we see that Nieuwentijdt's rhetorical device deployed in Christian apologetics became a part of science. The line between rhetoric and proposition had once again been thoroughly blurred. Secondly, while it is clear that Nieuwentijdt was engaged in a highly speculative exercise promiscuously manipulating orders of magnitude, the result of these numerical manipulations soon became absorbed into natural philosophy texts as a legitimate calculation. My next example underscores both of these points, but especially the first.

7.3.3 *The metaphor from ballistics*

While the sand metaphor connoted the smallness and superabundance of the light particles, another metaphor was employed to indicate their great speed. The speed of light was first measured by

Olaf Römer in 1675 and within a few years the Dutch physicist Christiaan Huygens had compared Römer's calculation with the velocity of a ball discharged from a cannon. According to Mersenne the cannon ball travelled at 589½ feet per second; hence such a ball would occupy a period of twenty-five years in travelling from the sun to the earth, a journey which light executes in some eight minutes. For Huygens, who supported a pulse theory of light, the comparison between light and a cannon ball was only appropriate in respect to their velocities, the ratio of which illustrated the immense difference between the speed of light and the highest speed achieved by human art. Indeed, for Huygens the great velocity of light constituted an argument *against* it being a material body.[24]

Huygens' calculation subsequently appeared in William Derham's Boyle lectures of 1711–12 which were published under the title *Physico-Theology*.[25] Here Derham recalculated the velocity of a cannon-ball – size unspecified – obtaining a smaller value than Mersenne, such that the ball would take thirty-two-and-a-half years to reach the sun. While Derham's recalculation further increased the disparity between the velocity of light and that of a bullet, his concern was less with the numbers *per se* than with surprising the reader by the immense *difference* between the two speeds. Derham's purpose in overawing the reader was apologetic since, like Nieuwentijdt, he sought to demonstrate God's providence. He described light as 'that noble and useful Creature of God' and dilated on its beneficial influence which reached man from the sun in such a small period of time. To emphasise God's providence, Derham contrasted the rapidity of light travelling with the 'Motion of the swiftest Bodies on Earth, such as a Bullet out of a great Gun', which would occupy over thirty-two years in passing from the earth to the sun.

Throughout the eighteenth century this metaphor derived from ballistics was utilised extensively in natural philosophy texts to convey the prodigious velocity of light. On occasions the comparison was merely stated, but frequently it was passed into rhetorical usage to contrast the power, wisdom and greatness of God with the feebleness of man. Thus, on being told the velocity of light, the pupil in Benjamin Martin's *Philosophical Grammar* (1735) asks:

Pray make this incredible Swiftness of *Light* a little more intelligible by Example.

[Tutor:] I will: 'Tis found that a Bullet, at his first Discharge from the Muzzle of a Cannon, flies one Mile in a little above 8½ Seconds, and therefore would be 32½ Years in arriving to the *Sun* . . . that is, *Light* flies one million five hundred and thirty thousand Times faster than a Bullet at his first Discharge. [Pupil:] Wonderful are the Works of God! Not only past finding out but past Belief, when but imperfectly known![26]

Equally dense in exclamation marks is the following passage from Adam Walker's *A System of Familiar Philosophy*:

With what inconceivable velocity must light impel light through such an amazing extent of space! ninety-five millions of miles in eight minutes! a million times swifter than a cannon ball! – the mind shrinks from the pursuit of such an idea.[27]

In both of these quotations, as in Derham's *Physico-Theology,* it is the vastness of the speed of light, the mind's inability to comprehend it, and the marvellous structure of God's providential plan that are emphasised. As a rhetorical device, the metaphor from ballistics stressed all these points, but it also drew the contrast between God's works and the puny products of human art and emphasised the limitations of the human mind.

Rhetoric, as we have often been warned, is seductive. The contrast with ballistics, as used by Derham and Martin early in the century for apologetic purposes, soon came to be replaced by the similarity with ballistics. The particles of light emitted from a luminous object were *like* balls shot from a cannon. However, to fill space with cannon-ball-like light particles had dire consequences. Soft, gentle light found itself pulled into the discourse of a heavy artillary bombardment. As early as the 1730s John Rowning rushed to the defence. The velocity of light, he noted, is more than a million times that of a cannon ball but, he added quickly, if these light particles were not so small, then 'the Eye would be rather wounded, than delighted by them' and tender flowers would be blown to pieces.[28] Other later writers stated the case more positively; the light particles must be inconceivably small, otherwise they would damage the eye by their great momentum.[29] The metaphor from ballistics had by the mid-century taken on a new lease of life. A metaphor intended to invoke surprise and wonder had come close to being taken as a description of the universe – with dangerous consequences!

This is the context in which we need to examine one of the most celebrated attacks on the projectile theory. In 1752 Benjamin Franklin wrote Cadwallader Colden a letter which was presented at the Royal Society four years later and published in 1769.[30] 'I must own', wrote Franklin, 'that I am much in the *Dark* about *Light.* I am not satisfy'd with the Doctrine that supposes Particles of Matter, call'd Light, continually driven off from the Sun's Surface, with a Swiftness so prodigious! Must not the smallest Particles conceivable, have, with such a Motion, a Force exceeding that of a 24 pounder discharg'd from a Cannon?' The question is a rhetorical one and Franklin was clearly engaged in a rhetorical game.[31] Franklin was further extending the widely-used metaphor from ballistics and by introducing the twenty-four pounder he made the comparison appear all the more realistic and all the more bizarre. Armed with this strategy Franklin turned the metaphor to his own purpose, which was to hold the projectile theory of light to ridicule and provide an *ad hominem* argument in favour of the vibration theory to which he subscribed.

Soon after its publication, Franklin's letter drew the fire of Bishop Samuel Horsley, the editor of the first collected edition of Newton's works. Horsley came out fighting in his paper, 'Difficulties in the Newtonian Theory of Light Considered and Removed', which was read before the Royal Society soon after the publication of Franklin's letter.[32] He admitted that the problem Franklin had raised is 'of some importance, and deserve[s] a strict [sic] discussion'. Franklin's rhetoric thus appears to have struck home and produced a serious response. But, just how strict and rigorous was Horsley's argument? Unlike Franklin he did indeed employ mathematics, but we must not be blinded by a few numbers. Instead we need to examine Horsley's premises, since although his argument might be valid, his premises were thoroughly gratuitous. In essence, all he needed to do to remove Franklin's objection was to guess at the density of the light particles and then to postulate their size with the view of obtaining a value for the momentum which would be small enough. In a sense, then, Horsley's use of mathematics, indeed his whole argument, was a rhetorical exercise to disarm Franklin's threatening objection. The metaphor from ballistics had worked against the projectile theory, so Horsley appointed himself to defend Newton by distancing light from heavy artillery. 'Dr. Franklin's . . . question is answered', concluded Bishop Horsley.

A particle of matter, which is probably larger than a particle of light, moving with the velocity of light, has a force of motion, which, instead of exceeding the force of a twenty-four pounder discharged from a cannon, is infinitely less than that of the smallest shot discharged from a pocket pistol, or less than any that art can create.[33]

An infinite conceptual distance is now imposed between light and the cannon ball. A similarity had once again been replaced by a contrast. The metaphor from ballistics had been stretched along another axis, or, more precisely, Horsley had protected the projectile theory by turning epiphor into diaphor.

7.3.4. *Weighing light*

The other main area where the materialist metaphor exhibited tension was in the progressive articulation of the projectile theory in the domains of research as well as teaching. Of the many problem areas which could be used to illustrate this point I shall confine discussion to one particularly interesting, although problematic, metaphorical extension of the projectile theory – the attempts made to detect the momentum of light.

If light is or is like projected matter then it too should display momentum, a property not usually associated with light. Throughout the eighteenth century a number of experiments were conducted to determine whether light has momentum. Before discussing these experiments I must introduce a note of caution. We tend to consider experiments as offering a means of testing scientific theories. In this particular case the experiment would be a crucial experiment to enable the scientist to choose between the projectile theory of light and some other theory, such as the wave theory, which predicted that the momentum would be zero. I have long felt uneasy about interpreting these momentum experiments in this way, partly because the several different attempts produced conflicting results which, incidentally, were consonant with the experimenter's theoretical preference – supporters of the projectile theory usually detected momentum, its opponents did not. But there is another problematic feature of these experimental reports and this concerns their style.

One of the first confirmations of the momentum of light was performed William Homberg and reported by Fontenelle in the *Histoire de l'Académie Royale des Sciences* for 1708.[34] Another frequently-cited confirmation was obtained by John Michell in the 1750s but was only published in Priestley's *History* of 1772.[35] What is striking in both accounts, like other similar ones, is their brevity and the perfunctory discussion of apparatus and experimental techniques. Moreover, both experiments were reported at second hand and Michell's only appeared in print some two decades after it was performed. Neither the experiments nor the experimenters spoke for themselves; instead, the reports of these experiments were widely displaced through space and time.

There are, of course, several hypotheses to account for the strange features of these so-called crucial experiments. In order to explore one such hypothesis we need to interpret the apparatus itself as the reification of a metaphor. In so doing we encounter a further use of metaphor for the historian of science, since metaphor is not confined to discourse or to texts but also takes on a materialised form in instruments. In the present case I will be concerned with metaphorical extension rather than with the larger problem of how scientific instruments incorporate metaphor. Each piece of apparatus is conventionally associated with certain specific functions, thus a balance is used for weighing and a convex lens is employed on an optical bench to increase the convergence of a beam of light. (It is, of course, also frequently used as a magnifying glass.) Metaphorical extension occurs when these items are employed in non-standard situations. Thus Homberg and Michell, who both accepted that light consists of matter particles in linear motion, projected the materialist metaphor onto their apparatus which was constructed out of standard items from the mechanics (not optics) laboratory. Homberg's apparatus consisted of a watch-spring fixed at one end with sunlight focused on the other end. Such a piece of equipment could be used to detect, say, the blow of a falling mass, but it was now being used in the domain of optics; significantly Priestley later referred to Homberg's apparatus as 'a kind of lever'.[36] Similarly Michell's apparatus was constructed from a length of harpsichord wire pivoted at its centre with a copper plate – onto which light was beamed – fixed at one end, and a counter-balance at the other. It was, in other words, a lever or miniature steelyard.

In 1799 Adam Walker reported a similar experiment using a 'small wire balance, nicely suspended on a point'.[37] Seven years earlier the *Philosophical Transactions* carried a long article by Abraham Bennet on the construction of a sensitive vibration magnetometer.[38] This paper included a brief description of an optical experiment in which Bennet tried to detect whether any impulse was conveyed to a delicately balanced illuminated screen. Bennet's experiment, like those of Homberg, Michell, Walker and others, was thus constituted by the materialist metaphor for light.

The results of these experiments were likewise expressed through metaphor. Homberg, for example, reported that his watch-spring was set into vibration by the incident light 'as if it had been struck by a stick'. Priestley and many other writers referred to these experiments as showing (or not showing) that rays of light possess *impulse* or *momentum,* and Priestley sought to measure this quantity in terms of weights (grains) multiplied by velocities; indeed, it is Priestley who talks of Homberg having proved 'the *weight* of the particles of light. This he [Fontenelle] advances not ironically, but very gravely'.[39] All of these terms used to describe the experimental results were drawn from mechanics but were being employed by metaphorical extension – more precisely by catachresis – in the domain of optics.

So, to return to my problem of why the experimental reports appear so uninformative. Firstly, in order to weigh light a small, delicately-balanced piece of apparatus is needed; not the kind of apparatus which could be displayed in a crowded lecture theatre. Moreover it requires bright sunlight, which would not be on tap and, even if available, could not readily be beamed into a lecture room. Hence this experiment forms no part of didactic practice. Secondly, while the question of whether light has momentum had become a question of scientific fact by the late nineteenth century it was treated as a matter of common sense in the eighteenth. In a didactic tradition in which there was little or no room for open-ended questions and where customers demanded a short, systematic course of instruction, the move from asserting that light is composed of material particles to claiming that it has momentum is one of continuous metaphorical extension. Metaphor works according to the interactionist theory by the transference of the qualities of the secondary subject on to the primary subject. Like the thief in the night, metaphor works silently; in this case convincing people that light has momentum. The experiments and experimental reports, then, take on a symbolic role, articulating the known rather than offering a crucial test.

7.4 *Conclusion*

There are several points worth drawing to the reader's attention about the role of metaphor in eighteenth-century optical texts. Firstly, metaphors – particularly the sand and ballistic metaphors –

are highly malleable, being subject to a wide range of different uses by various writers on optics. Thus a specific metaphor (far from being confined to a particular, culturally-determined function) could be used to establish a variety of incompatible tensive relationships.

Secondly, metaphor provides an appropriate site for studying the broader cultural dimensions of science. The fluid, sand and ballistic metaphors provide examples of how the language of optics was impregnated by theological discourse and theological values. It has also been noted that the latter pair of metaphors were initially articulated for apologetic purposes, but soon became part of science.

Thirdly, metaphors can act silently and imperialistically. While Boyd has been concerned with the way metaphors guide and encourage research, the example of momentum experiments suggests a further role for metaphor in connection with cognitive psychology since metaphor may lead scientists to presume the structure of nature. Moreover, as well as the naming function discussed by Boyd, I would also suggest a role for metaphor in the construction and description of experiments.

Finally, I want to argue that two common distinctions do not stand up to close scrutiny. The distinction between the teaching and research roles for metaphor seems less than well founded in the eighteenth-century material examined above; there seems to be no sharp dividing line between didactic and research texts. Indeed, the metaphor from ballistics inhabited both the popular lectures of Benjamin Martin and Samuel Horsley's defence of the projectile theory before the Royal Society. While this distinction has some uses, the position towards which I am moving makes metaphor *constitutive* of scientific discourse.

The other dichotomy which this study challenges is the one with which we began; the static opposition between the figural and the literal. What this study has shown is that, at least in the period and domain studied, metaphor played several non-trivial roles in scientific discourse. Nietzsche seems closer to the mark than Locke when he argued that all language is metaphorical, although some metaphors are more novel and active than others. Even knowledge, claims Nietzsche, 'is nothing but working with the favourite metaphors, an imitation which is no longer felt to be an imitation'. Meanings are constantly changing and it is this flux which for Nietzsche is the source of poetry and of all human creativity.[40] Viewed in this way eighteenth-century optics, so often portrayed as sterile, is seen to exhibit a creative, poetic dimension.

Notes

1 John Locke, *An Essay Concerning Human Understanding,* ed. J. W. Yolton, 2 vols., London and New York, 1961, I, pp. 105–6.

2 M. Johnson, ed., *Philosophical Perspectives on Metaphor,* Minnesota, 1981, p. ix. See also A. Ortony, ed., *Metaphor and Thought,* Cambridge, 1979; S. Sacks, ed., *On Metaphor,* Chicago, 1979; D. S. Miall, ed., *Metaphor: Problems and Perspectives,* Brighton, 1982; W. H. Leatherhead, *The Role of Analogy, Model and Metaphor in Science,* Amsterdam, 1974; M. B. Hesse, *Models and Analogies in Science,* Notre Dame, 1966.

3 R. M. Young, 'Darwin's metaphor: does nature select?', *The Monist,* LV, 1971, pp. 442–503; L. Rather, 'On the source and development of metaphorical language in the history of Western medicine', in L. G. Stevenson, ed., *A Celebration of Medical History: The Fiftieth Anniversary of the Johns Hopkins Medical School and the Welch Medical Library,* Baltimore and London, 1982, pp. 135–56.

4 S. C. Pepper, *World Hypotheses,* Berkeley, 1942; R. S. Jones, *Physics as Metaphor,* London, 1983.

5 I. A. Richards, *Philosophy of Rhetoric,* Oxford, 1936; M. Black, 'Metaphor', *Proceedings of the Aristotelian Society,* LV, 1954–5, pp. 273–94 (reprinted in Johnson, ed., *Philosophical Perspectives,* pp. 63–82); 'More about metaphor', *Dialectica,* XXXI, 1977, pp. 431–57.

6 Cited in E. Burtt, *The Metaphysical Foundations of Modern Physical Science,* Garden City, 1954, p. 202; R. Olson, *Science as Metaphor: The Historical Role of Scientific Theories in Forming Western Culture,* Belmont, 1971, p. 60.

7 The first two of these functions are incisively discussed by Richard Boyd, 'Metaphor and theory change: what is "metaphor" a metaphor for?', in Ortony, ed., *Metaphor and Thought,* pp. 356–407. See also Hesse, *Models,* pp. 157–66, and J. Martin and R. Harré, 'Metaphor in science', in Miall, ed., *Metaphor,* pp. 89–105.

8 There appears to be no consensus over the relation between metaphor and analogy. Following some of the authors cited above I take analogy to be the set of formal statements relating a scientific model and an explanandum, and governed by the relations discussed by Hesse, *Models,* p. 86. In a live metaphor, on the other hand, there exists a tension between primary and secondary subjects and the potential to generate new sets of relationships.

9 O. Hannaway, *The Chemists and the Word: The Didactic Origins of Chemistry,* Baltimore and London, 1975.

10 T. S. Kuhn, 'Metaphor in science', in Ortony, ed., *Metaphor and Thought,* pp. 409–19.

11 W. Jones, *Physiological Disquisitions; or, Discourses on the Natural*

Philosophy of the Elements, London, 1801, p. 122; J. Lacy, *The Universal System: Or, a Mechanical Cause of all the Appearances and Movements of the Visible Heavens*, London, 1799, p. 1; P. D. Leslie, *A Physiological Inquiry into the Cause of Animal Heat: With Incidental Observations on Several Physiological and Chymical Questions, Connected with the Subject*, London, 1778, p. 120; F. O'Gallagher, *An Essay on the Investigation of the First Principles of Nature: Together with the Application thereof to Solve the Phaenomena of the Physical System*, 2 vols., Dublin, 1784–5, I, pp. 331–2.

12 G. N. Cantor, *Optics after Newton. Theories of Light in Britain and Ireland, 1704–1840*, Manchester, 1983; C. B. Wilde, 'Hutchinsonianism, natural philosophy and religious controversy in eighteenth-century Britain', *History of Science*, XVIII, 1980, pp. 1–24.

13 G. Adams, *Lectures on Natural and Experimental Philosophy Considered in its Present State of Improvement*, 5 vols., London, 1794, II, p. 146; D. Forbes, *A Letter to a Bishop, Concerning some Important Discoveries in Philosophy and Theology*, Edinburgh, 1736; G. Horne, 'Essays and thoughts on various subjects', *The Works of the Late Right Reverend George Horne, D. D. Lord Bishop of Norwich*, 4 vols., London, 1831, I, p. 67. For a discussion of the role of language in mediating between the material and spiritual realms see G. Cantor, 'Light and enlightenment: an exploration of mid-eighteenth-century modes of discourse,' in D. C. Lindberg and G. Cantor, *The Discourse of Light from the Middle Ages to the Enlightenment*, Los Angeles, 1985, pp. 67–106.

14 Hannaway, *Chemists*, pp. 1–72.

15 I. Newton, *Philosophiae Naturalis Principia Mathematica*, London, 1687, pp. 227–31.

16 I. Newton, *Optice: Sive de Reflectionibus, Refractionibus, Inflectionibus & Coloribus Lucis*, London, 1706, p. 315.

17 J. Harris, *Lexicon Technicum; or, an Universal English Dictionary of Arts and Sciences, Explaining not only the Terms of Art, but of the Arts Themselves*, 2 vols., London, 1706–10, II, art. 'Light', n.p.

18 J. T. Desaguliers, *Physico-Mechanical Lectures. Or, an Account of what is Explain'd and Demonstrated in a Course on Mechanical and Experimental Philosophy*, London, 1717, p. 42; J. Rohault, *Rohault's System of Natural Philosophy, Illustrated with Dr. Samuel Clarke's Notes Taken Mostly out of Sir Isaac Newton's Philosophy*, 2 vols., London, 1723, I, p. 201; G. Atwood, *An Analysis of a Course of Lectures on the Principles of Natural Philosophy, Read in the University of Cambridge*, London, 1784, p. 132; J. Ferguson, *Lectures on Select Subjects in Mechanics, Hydrostatics, Pneumatics and Optics*, London, 1764, p. 124.

19 C. Smart, *Jubilate Agno*, ed. W. H. Bond, London, 1954, p. 113; G. Berkeley, *Siris: A Chain of Philosophical Reflections and Inquiries Concerning the Virtues of Tar-Water and Divers other Subjects Connected Together*

and Arising One from Another, Dublin, 1744, pp. 69–85; D. V. Erdman, ed., *The Poetry and Prose of William Blake*, Garden City, 1970, pp. 468–9. See also D. D. Ault, *Visionary Physics: Blake's Response to Newton*, Chicago, 1974, and K. Raine, *Berkeley, Blake and the New Age*, Ipswich, 1975.

20 B. Nieuwentijdt, *The Religious Philosopher; or, the Right Use of Contemplating the Works of the Creator. Designed for the Conviction of Atheists and Infidels*, 3rd edn., 2 vols., London, 1724, I, pp. 454–60.

21 Nieuwentijdt, *Religious Philosopher*, pp. 409–10.

22 B. Martin, *The Philosophical Grammar; being a View of the Present State of Experimental Physiology or Natural Philosophy*, London, 1735, p. 70.

23 Ferguson, *Lectures*, p. 125; O. Goldsmith, *A Survey of Experimental Philosophy, Considered in its Present State of Improvement*, 2 vols., London, 1776, II, p. 268; W. Nicholson, *An Introduction to Natural Philosophy*, 4th edn., 2 vols., London, 1796, I, p. 257.

24 W. Derham, *Physico-Theology: Or, a Demonstration of the Being and Attributes of God, from his Works of Creation*, 6th edn., London, 1723, p. 281.

25 Derham, *Physico-Theology*, pp. 28–31.

26 Martin, *Philosophical Grammar*, pp. 59–60.

27 A. Walker, *A System of Familiar Philosophy: In Twelve Lectures*, 2 vols., London, 1802, II, p. 232.

28 J. Rowning, *A Compendious System of Natural Philosophy*, 2 vols., London, 1734–8, II, p. 6.

29 For example, T. Rutherforth, *A System of Natural Philosophy, being a Course of Lectures in Mechanics, Optics, Hydrostatics, and Astronomy; which are Read in St. Johns College Cambridge*, 2 vols., Cambridge, 1748, I, p. 219.

30 I. B. Cohen, ed., *Benjamin Franklin's Experiments; a New Edition of Franklin's Experiments and Observations on Electricity*, Cambridge, Mass., 1941, p. 325.

31 For Franklin to have demonstrated that the momentum of the light particle is greater than that of the cannon ball he would have had to determine four quantities, being the velocities and masses of the two bodies. Of these, only three were known but not the fourth, the mass of the light particle. Owing to the indeterminacy of this quantity, it is clear that we are not dealing with a demonstration, but with rhetoric.

32 S. Horsley, 'Difficulties in the Newtonian theory of light, considered and removed', *Philosophical Transactions of the Royal Society of London*, LX, 1770, pp. 417–60.

33 Horsley, 'Difficulties', pp. 419–20.

34 Fontenelle's report of Homberg's experiment appears in his *Histoire de l'Académie Royale des Sciences*, 1708, p. 21. For other analyses of the experiments conducted to detect the momentum of light see M. L. Schagrin,

'Early observations and calculations of light pressure', *American Journal of Physics,* XLII, 1974, pp. 927–40; J. Worrall, 'The pressure of light: the strange case of the vascillating "crucial experiment" ', *Studies in the History and Philosophy of Science,* XIII, 1982, pp. 133–71; Cantor, *Optics after Newton,* pp. 52–9.

35 J. Priestley, *The History and Present State of Discoveries Relating to Vision, Light, and Colours,* London, 1772, pp. 387–90.

36 Priestley, *History,* p. 385.

37 Walker, *Philosophy,* I, pp. 4–5.

38 A. Bennet, 'A new suspension of the magnetic needle, intended for the discovery of minute quantities of magnetic attraction, etc.', *Philosophical Transactions of the Royal Society of London,* LXXXII, 1792, pp. 81–98.

39 Priestley, *History,* p. 385.

40 P. Cantor, 'Friedrich Nietzsche: the use and abuse of metaphor', in Miall, ed. *Metaphor,* pp. 71–88.

8 *Michael Moriarty*

Figures of the unthinkable: Diderot's materialist metaphors

'The material world,' continued Dupin, 'abounds with very strict analogies to the immaterial; and thus some colour of truth has been given to the rhetorical dogma, that metaphor, or simile, may be made to strengthen an argument, as well as to embellish a description.'

Edgar Allan Poe, 'The Purloined Letter'

I want to begin with a text of Fontenelle's that shows how the antithesis between science and rhetoric can both confirm and undermine itself, since it is this double process of undermining and confirming that I shall be looking at here in connection with Diderot. The 'Digression sur les anciens et les modernes' begins as follows:

Properly understood, the whole question of the superiority of the Ancients or the Moderns comes down to whether the trees that used to grow in our countryside were taller than those of today.[1]

Fontenelle argues that to conceive the ancients as naturally superior to us is as gratuitous as to imagine that the trees were higher then. It is all very well to claim that 'Nature exhausted herself in bringing forth these great Originals' (Homer, Plato, Demosthenes, and so forth), as if they belonged to a different species: but 'physical science is not in agreement with all these fine-sounding phrases'.[2] Already the antithesis, then; and it recurs when Fontenelle defends his choice of mode of argumentation: he could have resorted to the familiar rhetorical topoi of this controversy, but instead, 'it seemed to me that the quickest way of settling it would be to see what physical science has to say, since it has the knack of cutting short a good many disputes that rhetoric would prolong for ever.'[3]

The discourse of rhetoric is thus displaced by that of science; but so far from *leading to* a solution of the quarrel, the displacement actually brings it about in advance. For to dismiss rhetoric is to dismiss the field in which the ancient world can make its strongest claim to superiority; while in physical science the modern world can easily claim the advantage. Fontenelle's discursive move thus enacts what it purports merely to place on the agenda: the superiority of the moderns over the ancients. It is, of course, quintessentially rhetorical, resting on the metaphorical equation of a human being with a tree; doubly metaphorical, perhaps, in that not only is it not literally true, but also in that it tends to obscure what it does not throw into relief: the cultural, as the specific differentiation of the human from the animal world. It is true that Fontenelle goes on to argue that, since human beings are naturally the same in more or less all times and places, what differentiates one society or epoch from another must be 'external circumstances, such as time, forms of government, the state of public affairs'.[4] This clearly makes culture, in the broadest sense, the decisive factor, and shows that the metaphorical equation of human beings with trees is not, as it may appear to be, a collapsing of the cultural into the physical, but rather a rhetorical blow against a pseudo-Nature, a particular mystificatory discourse about the ancients' essential superiority. I want to read Diderot's materialist metaphors in much the same sort of way: not as reflections of, or pointers to, scientific truth, but as moves enacting the 'truths' they assert, dislodging a set of illusions.

To take up this position is to underline Lenin's contention that 'arguments and syllogisms alone do not suffice to refute idealism, and . . . here it is not a question for theoretical argument', a standpoint with which he specifically associates Diderot.[5] If the materialist theoretical project can no longer so confidently assert its self-constitution as science by the break from idealist discourses (ideologies, rhetorics), then materialism arguably needs to develop a theory of rhetoric and of its own rhetorics.[6] It is hoped that the present paper says something worth saying 'about Diderot', but also that it may in some sense contribute to this task: the task pointed to by Lenin, but which tends to disappear under the frequent crudity of his own rhetoric.

I shall argue here that Diderot calls into question the logic, the 'syllogisms and arguments' of anti-materialism by representing them as materially conditioned. To this end, he is obliged to resort freely to metaphor. In a literal sense, he writes off the argument that materialism is unthinkable by showing that thinking is metaphorical, and hence materialist, and by constructing images for thinking materialism.

To point to the impressive pedigree of the anti-materialist argument (Descartes, Pascal, Spinoza, and to some extent Locke) is at the same time to point to a relative historical unthinkability in the notion of thinking matter. I want to glance briefly at the arguments by which these thinkers refute materialism.

Descartes is criticised by Hobbes and Gassendi for simply assuming that the subject of thinking cannot be a body. In the *Sixth Meditation* Descartes has convinced himself that his clear and distinct perception of himself as a thinking thing excludes extension; but in his reply to Hobbes he does not scruple to extend this into a general assertion of the incompatibility of thought and extension.[7]

Pascal argues, for once, along quasi-Cartesian lines:

> I can certainly conceive of a man without hands, feet, or head, since only experience teaches us that the head is more necessary than the feet. But I cannot conceive of man without the power of thought. He would be a stone or a brute beast.[8]

It could of course be objected to this that there are problems in asserting the internal coherence of a concept ('headless thinking man') that experience invalidates; just as it could be objected to the analogous argument: 'The philosophers who tamed their passions: what kind of matter could have done that?',[9] that it presupposes what it purports to prove: the impossibility of matter acting self-reflexively, and thus the impossibility of accounting for human beings in purely material terms. But in any case there is a certain ideological profit in this incapacity of Pascal's to conceive of matter thinking. The immateriality of the soul, which the above fragment is intended to prove, is not here a staging-post in a metaphysical apology for theism; but it is proof of the double nature of man, of the greatness that must be installed in a dialectical relationship with the wretchedness, for the doctrine of the Fall to appear as the only believable solution to the enigmas of the human condition.

Spinoza urges against materialism, an objection akin to the above observations on Pascal: namely, that it asserts coherences that are purely illusory. His own system of course by-passes the dualist/materialist opposition by asserting that body and mind, substance thinking and substance extended, 'are one and the same substance, comprehended now through one attribute, now through another'.[40] But this is by no means tantamount to conceiving thought as a material process: on the contrary. The notion of a corporeal soul falls into the category of fictitious ideas: it comes from running together 'some corporeal image' with the *word* 'soul'. 'When we know the nature of the soul, we cannot imagine it as square' (or presumably as possessing any extension), 'though anything may be expressed verbally'.[11] In the last phrase, Spinoza points to language, conceived as the union of words with ideas, as a source of error, a theme developed elsewhere in the essay 'On the improvement of the understanding':

> Since words are a part of the imagination – that is, since we form many conceptions in accordance with confused arrangements of words in the memory, dependent on particular bodily conditions, – there is no doubt that words may, equally with the imagination, be the cause of many and great errors, unless we keep strictly on our guard. . . . Many things we affirm and deny, because the nature of words allows us to do so, though the nature of things does not.[12]

Diderot's concerns may be seen both as intersecting with and as responding to Spinoza's line of argument here. The common emphasis is on the determination of thinking by bodily conditions. What Diderot rejects is the possibility of externality to these conditions, of an escape from their influence into the space of the pure concepts of the understanding. Rather, his writing shows an engagement with, a commitment to language and the imagination, worked out in the instances of metaphor.

Before I proceed to argue this point, I shall mention a final witness to the unthinkability of materialism, to wit, Locke. True, he holds that we cannot know that matter cannot think, or that God could not confer on it the power of thought;[13] but this only makes more significant the horizons of conceivability on which he insists:

> If we suppose only Matter and Motion first, or eternal; *Thought* can never begin to be. For it is impossible to conceive that Matter either with or without Motion could have originally in and from it self Sense, Perception, and Knowledge, as is evident from hence, that then Sense, Perception, and Knowledge must be a property externally inseparable from Matter and every particle of it.[14]

If Locke, then, does not deny that matter might be able to think, he does deny that it can originate thought, perception, and sensibility. The *Rêve de d'Alembert* attempts to show that Locke's inconceivable is in fact possible.

But materialism has more to contend with than an established and prestigious philosophical discourse. The ideological pressure of dualism is more insidious than that. Diderot writes in the *Éléments de Physiologie* that the clarity of the materialist explanation of mind is obscured by 'the *almost innate* propensity [my emphasis] to postulate a superfluous entity as the judge of coexisting sensations, when all that is necessary is the sentient being itself which experiences them and gives them utterance'.[15] Paradoxically, the clear idea is one which we have a natural penchant to disbelieve. This penchant is not, of course, 'natural'; I shall argue that it is rather the effect of a certain ideological imaginary that Diderot seeks to disperse by the employment and transgression of metaphors. It could be said that in so doing he is continuing along the lines of Gassendi's objections to Descartes. Whether or not Gassendi is deliberately mobilising rhetorical devices against Cartesian metaphysics, this is certainly what Descartes perceives him as doing.[16] In effect, Gassendi implies that Descartes is thinking his anti-materialism with a crude image of matter,[17] that he has not taken seriously the images of the soul as material (a wind, a flame) which he mentions in the Second Meditation.[18] Descartes argues that the image one has of body (as 'subtle' or 'crude') makes no difference to the question of the mind's independence from it.[19] This is what Diderot's practice of writing contests.

The association of metaphor and materialism in Diderot goes back at least to the *Lettre sur les Aveugles*, a materialist work on two levels, in that it both offers a materialist cosmogony, an account of the world in general in terms of matter and motion, and seeks to explain specifically human phenomena, such as morality, in terms of an experience structured and determined as to its content by its physical co-ordinates. The blind man Diderot portrays has a peculiar horror of theft, to which he is especially vulnerable; by the same token, Diderot suggests, a society of blind people would either hold wives in common (one partner being much like another) or else punish adultery severely.[20] But in order to substantiate the connection between blindness and moral or metaphysical ideas, Diderot needs, not only illustrations taken from actual blind people, but an account of the determinate experience of blindness. To this, metaphor is indispensable.

The *Lettre sur les Aveugles* is itself a metaphor from the outset, the substitution of one set of sensory perceptions for another. The opening pages of the letter present it as a compensation for Diderot's inability to take his correspondent to see Réaumur removing a cataract. The scientist's refusal to allow an audience to be present at the operation is justified as follows: 'The observations of so famous a man stand less in need of *spectators* while they are being carried out than of *hearers* when they have been carried out'.[21] Spectacle has to be transformed into discourse: immediate presence has no privileged cognitive value. (For similar reasons, Diderot casts doubt on the value of attempts to solve Molyneux's problem (would a blind person who could distinguish shapes by touch be able, on recovering his or her sight, to distinguish the same shapes simply by looking?) by questioning people who have just had a cataract removed. The only patient who could give a useful account of his or her sensations would be one with the philosophical competence to compare the prior experience of blindness with the subsequent experience of sight.)[22] The *Lettre* substitutes discourse for immediate experience, discourse at the second degree, moreoever, since it is largely made up of reported conversation (Diderot collapses these two levels by also terming it an 'entretien'.)[23]

Itself a substitution, then, the *Lettre* sets in motion two chains of substitutions; the expedients by which the blind compensate for their deprivation, and the verbal expedients, the catachreses, by which sight is presented in terms of other sensory activities. The blind man of Le Puiseaux, interviewed by Diderot and friends, constantly struggles to make sense to himself of the sense he does not have in terms of those he does. His success in doing so is measurable by his normal auditor. At the same time, the workings of the metaphor help to reveal general problems of the language of sensory perception in which the normal speaker or writer is also involved.

The world of the blind is one great set of substitutions, concrete catachreses (metaphors without literal equivalents). For the man of Le Puiseaux, night is day, for it is by night that he lives his waking life, when nothing is in his way and he in no-one else's;[24] and Melanie de Salignac, the blind girl described by Diderot in the *Additions à la Lettre sur les Aveugles*, used to say, as night drew on, that the reign of the normal was nearing its end, while hers was about to begin.[25] But the crucial substitutions take place at the linguistic level. The blind man, asked to define a mirror, replies that it is 'a machine . . . that places objects in relief at a distance from themselves, if they happen to be in the proper spatial relation to it'.[26]

Diderot argues that this definition necessarily follows from the subject's experience. Other people apprehend objects by sight, as he does by touch; but he knows, further, that though you can touch your own face, you cannot see it. Sight, then, is a kind of touch that applies to objects other than one's own face, at a distance. Since a mirror reproduces objects away from themselves, and the blind man can conceive of physical existence only in terms of relief, he concludes that the mirror-image is a sort of tactile dummy of the object, at a distance from it, and is surprised to learn that it is not in fact tangible.[27]

The metaphor, sight as touch, has broken down here. But although it misleads its user, it is still subject to the control of the normal reader or hearer, who can measure the blind man's concept of sight against the experienced reality, who can realise that the blind man necessarily confuses two aspects of sight: he rightly conceives it as – what touch is for him – a sensory input enabling purposive physical activity, but cannot apprehend its specifically optical nature. In other words, he understands sight figuratively, but not literally: that is, catachrestically.

The catachresis inevitably works its way into Diderot's language. Thus he says, of the eyeless Cambridge mathematician Saunderson, whose tactile sensitivity was so acute that he could distinguish the least variation in the atmosphere, that 'he saw . . . with his skin'.[28] In point of fact, although the normal can flatter themselves with the sense of their own superiority (superiority of sense), their position vis-à-vis the language of physiology is uncomfortably close to that of the blind man. When asked to define the eye, the latter says that it is 'an organ on which the air has the same effect as my stick does on my hand;[29] and he extends the comparison by pointing out that when he touches the wrong object with his stick, it is like when an object interposes between the eye and another object. The blind man has been astonished when the different varieties of mirror (enlarging, distorting, and so on) have been described to him; now, when they hear this definition of the eye, it is his listeners' turn to be surprised. The blind man's account of vision exactly corresponds to that of Descartes, which no-one, says Diderot, has ever subsequently bettered: 'Open Descartes's *Dioptrics*, and you will see the phenomena of sight related to those of touch, and optics illustrated by picture after picture of men engaged in *seeing with sticks*'.[30] Descartes himself has to resort to catachresis, in the broader sense of a metaphor in which words are perverted from their literal meaning: just as those born blind are forced to rely on a stick, he is forced to say that 'they see with their hands'; forced by the very nature of light itself:

Light is nothing other, in those bodies that are called luminous, than a certain movement, or a very swift and intense activity, that goes towards our eyes via the air and other transparent bodies, just as the movement or the resistance of the bodies encountered by the blind man goes towards his hand via his stick.[31]

The truth of sight does not reside in the intuitions produced by its presence; sight cannot be spoken in its own terms. Rather, it is expressible only through a figure of speech derived from the situation of its absence. As regards language, Diderot's blind man and the theorist of optics are in the same position.

It might of course be objected that Descartes is only resorting to the image of the blind man's stick the better to explain a state of affairs already understood quite independently of it, simply using comparisons 'which help us to conceive [light] in the manner that I believe is most suitable for explaining those of its properties that we can apprehend by experience and then deducing all those other properties that we cannot observe so readily'.[32] But it is obvious from this very passage that the comparison is no mere communicational expedient but a cognitive instrument for interpreting experimental data and drawing deductions from them; the comparisons are compared to the suppositions of astronomers, which, though false in themselves, are yet related to actual observations and can lead to trustworthy conclusions.

If blindness thus metaphorically produces the truth of sight, it is not, by its necessary recourse to metaphor, condemned to untruth. 'Normality' can only construct itself by reference to the abnormal. Indeed, by a curious reversal, the blind person becomes the norm, dependence on touch coming to appear as paradigmatic of human beings' relation to their senses. In the *Entretiens sur le Fils naturel*, Diderot asserts that all the arts are a form of imitation, just as all the senses are a form of touch, and the same position is taken up in the *Rêve de d'Alembert*.[33] The blind person's touch is less a kind of second sight than sight itself is a form of second touch. To erect touch, one of the more 'physical' senses, into the paradigm is to displace, in a materialist gesture, the philosophical tradition that identifies sight as the privileged, because the most 'spiritual' sense.[34] Diderot goes further towards subverting the privileged position of normalcy: speculation, he says, depends on abstraction, and abstraction involves the ability to distinguish the sensible qualities of objects from one another, and from their material support; the blind person's imagination, which recollects and combines sensations of palpable points, is more abstract than that of the normal person, and the blind are therefore less likely to err in speculation.[35] Touch thus paradoxically becomes more abstract than sight, and the intellectual (since touch retains its earthy associations) is thus disjoined from the spiritual.

The assumed superiority of normalcy is again called into question by the *Lettre*'s discussion of metaphor. Saunderson is said to have been 'fertile in felicitous expressions', which Diderot glosses as expressions that are literally coherent in terms of one of the (physical) senses, and figuratively so in terms of another. They thus produce a double enlightenment, directly through the literal sense, and by reflection (Diderot's term) through the metaphor.[36]

But of course the unwitting user of the metaphor receives only the partial light cast by the direct sense; even someone as intelligent as Saunderson could perceive only half the ideas attached to the terms he was using. But we are all, Diderot insists, at times in the same position, despite our normalcy. If Saunderson's blindness prevents his recognising a metaphor for what it is, it is nevertheless what causes the metaphor to work. He is like the foreigner, whom paucity of vocabulary obliges to make the most of what linguistic resources he or she actually has; again, sometimes to happy effect. And these deviants, the blind person and the foreigner, are no differently placed from the genius, who is forced to deviate from common speech because it contains insufficient literal expressions to meet the needs of his or her exceptional imagination (Diderot cites Marivaux and Tacitus in this connexion).[37] Metaphor is thus the point of departure for a move that undermines the idealist identification of quantitative norm (having five senses; conforming to majority linguistic usage) with qualitative value, by foregrounding the qualitative effects produced by positions outside the norm: blindness or genius. This suggests an equation between the genius, with his or her 'lively imagination', and the blind, as physically determined individuals; and while it does not square exactly with the materialist theory of genius propounded in the *Rêve de d'Alembert* and the *Paradoxe sur le Comédien*, it certainly looks forward to it.[38] Saunderson, whose unwitting metaphors delight his hearers, is in the same position as the actor of the *Paradoxe*, who, himself unmoved, moves the audience to tears.

In other words, the argumentation of this part of the *Lettre* has tended to establish two materialist positions. Distinctions between physical normality and abnormality are pulled out of any secure relation with distinctions of value, thus calling into question the very terms 'normal' and 'abnormal', and along with them a certain teleology. Secondly, the sensory configuration of the individual has been shown to condition a whole range of supposedly pure intellectual activity, metaphysical, linguistic, and aesthetic, as well as moral ideas and attitudes. These two themes are developed into the basis of a materialist cosmogony in the second part of the letter. Firstly, Saunderson's blindness is represented as making him insensible to the theist's appeal to the harmonious spectacle of nature; secondly, his sensory privation is presented as evidence against the existence of a benevolent creator and in favour of his own deathbed vision (if the word is appropriate) of a succession of crippled worlds and monstrous creatures giving way gradually to viable systems and species.

The power of the *Lettre* as a materialist treatise comes largely from its taking metaphors literally. By establishing a rigorous homology between the role of touch in the blind person's life and that of sight in the normal person's, Diderot produces a radical challenge to theological metaphysics.

The *Lettre sur les Sourds et Muets* shows Diderot again at grips with the language of sensory deprivation and its relation to metaphor. Here again we are faced with a subject, deaf and dumb from birth, whose metaphorical performances call into question the certitudes of philosophy. Like the blind person, Diderot's deaf-mute is obliged to communicate by means of the categories furnished by his own limited sensory experience. Again, however, his difference from the normal person cannot automatically be seen as pure inferiority. Like the blind man of Le Puiseaux, he can astonish those around him, sometimes by an extraordinarily apt expression, which he himself may not appreciate, sometimes by a striking misconception that is none the less illuminating because clearly determined by the limitations of his own experience. In the first category, we may place the pantomimic gestures by which he expresses his contempt when Diderot accepts the bystanders' help with a chess problem. Not only does he make himself perfectly clear; his gestures actually remind several of the onlookers of the expression 'le tiers, le quart et les passants' (every Tom, Dick and Harry), which he himself can never, of course, have encountered.[39] This is a linguistic felicity comparable to those defined in the *Lettre sur les Aveugles*, literal with respect to one of the five senses, metaphorical with respect to another: for the deaf-mute has communicated his meaning visually, as he intended (by gesture), but has also unwittingly conveyed a verbal (therefore in this context, an auditory) message.

But he is not always so happy. Sometimes the ideas he succeeds in communicating can be shown to be false from the standpoint of a normal person. Thus, Diderot takes him to see Père Castel's 'ocular harpsichord', a device with an arrangement of coloured tapes, one for each finger-key, through which light passed, to produce a synthesis of musical and visual sensations. The deaf-mute cannot of course appreciate the musical part of the experience, but his interpretation of it is remarkable. He imagines that the harpsichord's inventor is a deaf-mute like himself, who is using the device to communicate with other people, and that each shade of colour stands for a letter of the alphabet; the colour-sequences formed by playing on the keys thus convey words, sentences, whole speeches. Not content with this, the deaf-mute goes on to conclude that music is a mode of communicating thought, and the various instruments so many organs of speech.[40]

The levels of metaphor operative in this passage are numerous. In the first place, the ocular harpsichord is itself a kind of sensory-metaphor-machine, for it transposes acoustic (musical) into visual stimuli. The deaf-mute's account of it does justice to its metaphorical nature (he sees it as representing letters by colours), while misreading the terms of the metaphorical transformation. When he concludes that music is a mode of communicating thought parallel to language, indeed parasitic on it, he is again mistaken; yet not entirely. For Diderot states that 'if sounds do not depict thought as clearly as speech, they do none the less say something'. Norman Rudich draws attention to the parallels between this passage and others in the *Entretiens sur le Fils Naturel*.[41] The first of these assigns the orchestra the role of *speaking*: rendering speech and imitating action; the second, already mentioned above, brings all the arts together under the rubric of imitation.[42] In other words, the deaf-mute's misperception retains a metaphorical validity as a pointer to the aesthetic

Diderot is seeking to construct. But this is not all. Again Diderot calls into question, even while acknowledging, the superior position of the normal person. The deaf-mute's account of music is manifestly false from the normal point of view, but from his own it is inevitable – in Diderot's words, 'a virtual logical certainty'.[43] The position is similar with the suppositious deaf and blind philosopher, who, Diderot suggests, would locate the soul in the tips of the fingers,[44] and the parallel is even closer with the blind man's erroneous definition of a mirror mentioned above. This and the deaf-mute's account of music are both compared with the misleading systems of philosophers; favourably so, since at least they result from the rigorous interrogation of experiential data, necessarily limited though these may be.[45]

The implications of the comparison are plain. The blind man and the deaf-mute offer instances of systematic misrecognition, determined by a specific set of sensory capacities. It so happens that we have, in the normal person's experience, both a criterion by which the accuracy of their thinking can be judged and the explanation of their errors. But these safeguards are lacking in the case of human philosophical reflexion in general. This too, it is suggested, is determined, and for all we know distorted, by a peculiar configuration of senses regarded as normal because statistically the most frequent. Philosophers have erred in making judgements of objects the nature of which cannot, or at least may not, be apprehended by our actual sensory equipment. That a hypothetical state of things is inconceivable to us, as music is inconceivable to the deaf-mute, is in itself no reason for asserting its impossibility: its internal coherence may simply be indiscernible to our limited and physically determined reason.

The particular type of catachresis by which the operations of one physical sense are described in terms of another thus plays an important role in damaging the security of the human philosophical subject, conscious of employing his rationality to attain eternal and essential truth.[46] It suggests that the apparent inconceivability of materialism should be judged less in terms of our perception of the internal and essential compatibility or otherwise of matter and sensibility, matter and self-movement, than with regard to the physically determined sensory equipment that provides the material for our abstract speculations.[47] Thus, when d'Alembert suggests, in the first of the conversations of the *Rêve*, that matter and sensibility are essentially incompatible, Diderot asks how he can know this, 'when you know the essence of nothing whatever, either matter or sensibility'.[48]

But though this line of argument excludes the *a priori* dismissal of materialism that we have encountered in Pascal, Descartes, Spinoza (to some extent) and Locke (to some extent), it does not provide positive grounds for thinking materialism a tenable, let alone the most cogent, account of the universe. In order to establish this later position, Diderot has to resort again to metaphor, but in a different way.

An example of this procedure may be found elsewhere in the *Lettre sur les Sourds et Muets*. The major explicit concern of the work is the problem of inversions of word-order in Latin, Greek and French. Diderot notes that the first word of a Latin period, expressing the idea that the speaker wishes to present first of all to the listener, may be grammatically modified by a later word by which it is governed. For the first word to receive the correct inflection, it is necessary that the idea conveyed by the grammatically dominant word should be present to the speaker's mind at the same time as the idea to which he or she wishes to give expressive priority; in other words, he or she must be able to think of more than one thing at a time.[49] But if the mind is a unity, as the conception of it as a spiritual substance implies, how is this possible?

Diderot's answer takes the form of a metaphor, that of the human clock, that appears to render the dualist hypothesis superfluous. The clock-image had earlier been applied to human beings by La Mettrie, transgressing the Cartesian schema that confines it to animals and the human body, but Diderot's treatment of it is particularly systematic.[50] In his hands, it becomes a conceit, with multiple correspondences between vehicle and tenor. The heart is the mainspring, and the organs housed in the chest are the other main elements in the movement. In the head, Diderot imagines a bell, fitted with little hammers; to these are attached an infinite number of strings terminating in all the various parts of the case (the limitations of the vehicle are closely being transgressed already). Above the bell is a figurine of the type found on clocks, its ear cocked like that of a musician checking that his instrument is in tune. This is the soul, although it is clear that 'soul' here is merely a courtesy title. Imagine then that several of the strings are pulled simultaneously: the bell will be struck several times, and the figurine will hear several sounds at once. If one imagines some of the strings as being pulled continuously, their sound will correspond to the sensations within us, such as that of our existence, that escape attention by their very continuity.[51]

Diderot takes the comparison further. The sounds of the bell do not die away at once: they go on, they form chords with their successors. The figurine can hear these chords, and judges them consonant or dissonant. Present memory (*mémoire actuelle*) consists in the resonance of the bell, judgment in the formation of chords, and speech in their succession. Diderot works the analogy even harder to elicit the nature of the thinking-process, and in so doing indicates the importance he attaches to musical patterns of thinking: there is a rule, he points out, that in harmonic passages a chord and its successor should have at least one sound in common, and this sound he compares with the middle term of the syllogism.[52]

Diderot's sudden abandonment of this comparison is a point I shall return to. But the clock-image inevitably calls to mind, if only by force of contrast, the more celebrated image that occurs in the *Rêve de d'Alembert*, of the human being as a '*clavecin sensible*', a sentient harpsichord. Before examining this image, however, I wish to glance at the enabling fiction of the dream itself, and its relation to metaphorical discourse.

The importance of metaphor in the *Rêve*, as an intellectual productive force and not merely a stylistic ornament, has of course been noted before. Herbert Dieckmann sees metaphor as the symbolic expression of the work's dominant principle – metamorphosis – and as 'the means par excellence in the refutation of dualistic, abstract and rigidly categorical thinking'.[53] The use of metaphor in the *Rêve*, as John Pappas likewise points out, reflects Diderot's view of analogy as 'a valid means of acquiring knowledge'.[54]

One of the key discussions of analogy occurs in paragraphs XXX and XXXI of *De l'Interpretation de la Nature*. The habit of experimentation can enable one to develop a nose (Diderot's expression) for remote analogies or resemblances in nature, and these can serve as the basis for empirically testable conjectures. But these conjectures are termed 'extravagances', ideas as wild and disjointed as the dreams of the sick.[55] This passage gestures not only backwards, to the materialist world-vision of the dying Saunderson (who then lapses into delirium),[56] but forwards, towards the scenario of the *Rêve de d'Alembert*.

It is clear that the valorisation of analogical thinking involves

Diderot in metaphorical writing, and clear also that the dream (the fiction of the dream) is going to become intertwined with a network (a very Diderotian image) of metaphors. The dream-vision in the text of philosophy goes back, of course, to antiquity (the vision of Er in the *Republic*, or of Scipio in Cicero's *Somnium Scipionis*); it reappears, with Descartes, at the birth of early-modern philosophy.[57] D'Alembert's dream, like these others, conveys a hidden truth requiring more or less interpretation, but it requires to be interpreted according to different codes. Ancient and medieval dream-theory distinguished dreams produced by natural (physical or psychological) causes (*somnia naturalia* or *animalia*) from those that conveyed a supernatural revelation (*somnia coelestia*). Only the latter category had any truth-value. Diderot, as it were, runs the two levels together: d'Alembert's dream is a revelation precisely because determined by his bodily condition and mental (that is, bodily) preoccupations. The authoritative figure that so often in classical or medieval dream-visions conveys truth to the dreamer is here replaced by the body itself, and the truth the body conveys is precisely that of the identity of its own substance with the substance of the universe as a whole. The dream thus enters into the self-referential, self-confirming pattern of the text of the *Rêve*.

For Bordeu in the *Rêve* and Diderot himself in the *Éléments de Physiologie* there are two types of dreams, the 'ascending dream', where a disturbance of the nerve filaments in a particular part of the body causes an appropriate image, visual or acoustic, to arise in the brain, and the 'descending dream', where an image arising in the brain produces a physiological reaction in the corresponding part of the body.[58] Thus, the image of a person one loves arising in the brain will generate a tension in the voluptuous filament, leading to orgasm. D'Alembert admits to having had such a dream the previous night,[59] and its genesis can be discerned from the notes, and through the imperceptions, of Mlle de l'Espinasse. The vision of a host of spontaneously generated animalculi and an ocean of heterogeneous molecules arouses a tension in the dreamer that seeks relief in the image, both comforting and stimulating, of Mlle de l'Espinasse, with the result envisaged by the physiologist.[60] The workings of d'Alembert's mind are governed by the law of the signifier: the theme of spontaneous generation persists in his drowsy suggestion that the semen he has just ejaculated should be bottled up and sent to the microscopist Needham ('discoverer' of spontaneous generation), while the vision of the ocean of matter leads to divagations on the reproduction of fishes. This confirms Diderot's assertion, in a letter to Sophie Volland of 20 October, 1760, that 'there is nothing disconnected either in the head of a person dreaming or in that of one insane', just as an apparently disjointed conversion is in fact held together by often imperceptible connections.[61] In general, one can say that the representation of d'Alembert's dream, and the dream-text, as determined by and determining the state of the body, enacts and confirms the materialism asserted by the *Rêve* as a whole.

D'Alembert's is not the only dream in the work: dreaming is dispersed, indeed disseminated, throughout. The dream is in fact a point of junction for philosophico-scientific and ethico-sexual materialism, and it is no accident that the *Suite de l'Entretien* should be a discussion of sexual morality.

For Bordeu, sleep is the breakdown of the whole personality, with the master (the central nervous system) abandoned to the discretion of his vassals, and to his own untrammelled activity.[62] This seems to be partially true of the state of waking reverie as well. It is not surprising that the relaxation of habitual controls, combined with the scope for autonomous activity on the part of the voluptuous filament, should produce associations between dreaming and the sexual. They are evident in d'Alembert's case, but they are also present in the reverie (Diderot's word) in which Mlle de l'Espinasse comes upon the 'completely mad idea' that 'man is perhaps nothing but an abnormal form of woman, or woman an abnormal form of man', and in Bordeu's meditation (again a *rêverie*) on d'Alembert's vision of human beings resolving into an infinity of human animalculi.[63] The breakdown of rational control in the dream is thus far from negative in its effects (though initially identified by Mlle de l'Espinasse as pathological): the moral internal censorship that blinds us to the pervasive influence of sexuality and the ideological internal censorship that shores up the traditional wisdom of philosophy are simultaneously, fruitfully, suspended. The materialist vision of the world is so seductive as even to overcome d'Alembert's rumoured impotence.[64] When Bordeau ironically assents to Mlle de l'Espinasse's exasperated assertion that there is no difference between a dreaming philosopher and a waking doctor, he affirms the characteristically Diderotian tenet that scientific experiment and observation are the necessary test and counterpart of analogical interpretations of nature.[65]

It is now time for the sentient harsichord. D'Alembert raises the question, similar to that posed in the *Lettre sur les Sourds et Muets*, of how we are able to think of two things at the same time, as we must be in order to utter the simplest proposition, since the proposition involves affirming or denying a quality of an object. Diderot replies:

I agree: which is why I have sometimes been led to compare the fibres of our organs to sentient vibrating strings. The sentient vibrating string oscillates and resonates for a long time after it has been plucked. This oscillation, this kind of necessary resonance, is what keeps the object present, while the understanding concerns itself with the quality that belongs to it.[66]

Moreover, the vibrating strings have the capacity to cause others to vibrate; in the same way, an initial idea can give rise to a second, these two to a third, and so on to infinity. The metaphor itself is a legitimation of metaphor, considered as the fusion of apparently remote ideas:

This instrument makes extraordinary leaps, and the arousing of an idea will sometimes cause a harmonic to vibrate that is separated from it by an incomprehensible interval. If this phenomenon is observed to take place between sounding strings that are lifeless and separated from each other, how could it not take place between living and connected points, between continuous and sentient fibres?[67]

D'Alembert is impressed; but he counters by suggesting that Diderot is falling into the very trap he is seeking to evade: he rejects the distinction between mind and body, yet he is treating the subject's understanding as an entity distinct from the instrument, a kind of musician that listens to the vibrating strings to pronounce on their consonance or dissonance.[68]

This figure is in fact akin to the 'incomprehensible miniature harpist' of the *Eléments de Physiologie*,[69] the metaphorical entity to which spiritualist thinking unconsciously has recourse. But it is also akin to the figurine of the *Lettre sur les Sourds et Muets*. D'Alembert's objection in fact raises enormous problems with the image of the human clock presented in the latter text. Its components are contradictory: the mechanistic-functional model is transgressed by the incorporation, or non-incorporation, of the sentient figurine, which is by nature, moreover, an ornament – as Diderot himself implicitly acknowledges.[70] In the *Rêve*, Diderot rebuts the allegation of closet dualism by stressing the difference between the literal harpsichord and the figurative human harpsichord. The latter is both musician and instrument, player and played. Similarly, in the *Éléments*, memory is referred to as a book that is its own reader.[71] But what we call a book is something that cannot read itself; a harpsichord, as such, requires a player distinct from it.

Diderot is thus obliged to cancel the image in the moment of affirming it. He is caught in a specifically materialist dilemma, which is, in effect, a metaphoric aporia. The imagery of vibrating strings, hammers and bells points, by itself, to a phenomenalist view of the subject as merely the locus of a string of sensations. The inclusion in the *Lettre sur les Sourds et Muets* of a component standing for the activity of consciousness (the figurine) threatens a relapse into dualism. In both cases, either a specific material component or the material system as a whole has to be vivified, surreptitiously in the *Lettre*, overtly in the *Rêve*. But this is not so satisfactory when the point at issue is precisely the conceivability of pure matter as alive and sentient. The mechanical-materialist metaphor cancels itself.

The problem recurs. The idea of thinking matter cannot simply remove or dissolve a prior limit of thinkability; it must always transgress it, just as the terms of the materialist metaphor must be wrenched from their literality. There is no simple sublation of the inanimate into the animate; the animisation of matter (the recognition of matter's animation) can only be the product of an intervention, a transformation, just as the marble statue of the *Entretien entre d'Alembert et Diderot* can only be sensibilised (proved to be inertly sensible) by a process of production.

Diderot, then, is not looking for the ultimate metaphor that will both express the essence and convey the truth of materialism. For this would imply the pre-existence of materialist truth, waiting only for Diderot to achieve its self-recognition. It would cancel the *Rêve*'s character as an intervention, as subversion. On the contrary, the dialogical form of the text implies that truth has to be actively appropriated, in fact produced, by each individual. Julie de l'Espinasse becomes a materialist (thus departing from an initial incomprehension that is also confidence in a certain ideological sanity, the sanity that enables her to brand as madness the dream of d'Alembert and the science of Bordeu) when she too begins to have wild ideas, to produce metaphors as extravagant as any of the metamorphoses dreamed of by d'Alembert.

Diderot's materialist metaphors cannot in fact be considered (as I have been considering them up to now, juxtaposing images from different works) outside of the textual economy they operate in; to separate them from this is, precisely, idealist. The dialogical economy of the *Rêve* both reflects and displaces Diderot's theses about analogical thinking. As was mentioned above, the analogy has to receive experimental verification. In the *Rêve*, this is represented by Bordeu, who stands as reality principle to the pleasure principle that fuels the conversational exchange ('it is a pleasure to talk with you', he says to Julie: 'Not only do you grasp what is said to you, but I am amazed by the soundness of the inferences you go on to draw from it').[72] He is 'the person who has practised medicine and carried out a certain number of observations'.[73] It is he that provides the empirical instances, the case-histories that serve to validate the wild analogies, that inscribe the metaphor within the text of scientific truth. D'Alembert toys with the idea of human polyps, but only in order to reject it; Bordeu instantiates it with the story of the Hungarian Siamese twins. A 'wild supposition' of the dreamer's is certified by him as 'the actual history'. L'Espinasse's 'completely mad idea' that 'man is perhaps nothing but an abnormal form of woman, or woman an abnormal form of man' is confirmed by the anatomical science in Bordeu's possession.[74]

But even Bordeu's access to science is not unmediated, or mediated only by observation and experiment. He too, as we have seen, has to submit to the dream-process, to the recognition of metamorphosis across 'an incomprehensible interval of centuries and successive developments', to letting his thoughts resonate across the 'incomprehensible interval' between the original note (if any note were original) and its harmonic: the incomprehensible interval of metaphor.[75]

The writer who specifically identifies Diderot as the precursor of his own twentieth-century materialism – Lenin – defines matter as 'an *objective reality* . . . existing outside our mind', and again as 'objective reality existing independently of the human mind and reflected by it'.[76] 'The antithesis of matter and mind', he says elsewhere, 'has absolute significance only within the bounds of a very limited field . . . the bounds of the fundamental epistemological problem of what is to be regarded as primary and what as secondary. Beyond these bounds, the relative character of the antithesis is indubitable'.[77] The impossibility and the necessity of materialist metaphor are inscribed within these two formulations. In the contradictions of the materialist metaphor, the shortfall between mechanical vehicle and human tenor, we can read the specific position of the human in the material universe, the relative separation between mind and matter. As Diderot says in his refutation of Helvétius, 'I am a human being, and I need distinctively human causes'.[78] But the derivative, secondary character of mind appears in the subject's dependence on material metaphors in order to speak its own nature and place in the universe: dependence, not mastery, since the material world cannot furnish the image that will finally express the place of human beings within it. Matter is heterogeneous, and the (material) subject's relation to matter is one of heterogeneity, since it cannot speak itself adequately through matter.[79]

The texture of the *Rêve* resembles matter in it heterogeneity. Its metaphorical tissue could be described – metaphorically – as the product of epigenesis rather than preformation. The different images are not homogenous animalculi sprung from Diderot's philosophical loins. Rather, the text has been formed by their successive apposition, like molecules. But this metaphor too must be cancelled as soon as put forward. For not only, if pressed, would it suggest a facile analogy between Diderot's biological views and the structure of his text – misleadingly, since though he vigorously rejected preformation, Diderot was not an out-and-out epigeneticist (Bordeu points to difficulties of epigenesis in the *Rêve*);[80] it would suggest a comforting organicism in the text, that the assertion of heterogeneity is precisely aimed at dispelling. The self-playing harpsichord has affinities with other images in Diderot (the self-reading book of the *Éléments*), but it cannot be placed on all fours with the other metaphors of the *Rêve* – the swarm of bees coalescing to form a single organism, the spider in the middle of its web who corresponds to the centre of the nervous system.[81] Bordeu, it is true, responds to L'Espinasse's production of the spider metaphor by equating it precisely with the swarm of bees mentioned earlier;[82] but these two images differ from that of the harpsichord inasmuch as they are taken from the living (as distinct from the potentially living) world. The problems touched on earlier are not present here in the same way, in that we are invited to think of the body as focused on a *living* material centre; even so, the vehicle of the metaphor has to be modified in keeping with the tenor as known from elsewhere, for we have to suppose that the web is actually a sensitive part of the spider.

In his essay on 'La mythologie blanche', Derrida draws attention to a strategy, in such different writers as Freud, Bergson and Lenin, of multiplying antagonistic metaphors with a view to neutralising or controlling their effect.[83] Along similar lines, Élisabeth de Fontenay thus describes Diderot's attempt to counter the danger pointed to by d'Alembert, of resurrecting the soul in the form of a *sensorium commune*, a register and judge of all sensations:

To prevent [the spiritual principle] from re-establishing itself, what is necessary is to find metaphors and models that will firmly maintain the fragmentation and dispersal [of the human being] into different agencies tugging in different directions, and that will permanently dissolve that cunning entity, the soul, always ready to lay claim to pride of place, under the pretext of kindly offering its centralising and highly undemocratic services.[84]

This strategy involves both commitment to and distrust of metaphor, an ambivalence that, one is tempted to say, goes well with, reflects the paradoxical position of human beings in the material universe.

One is tempted to say . . . but the temptation should be, for a little longer at any rate, resisted. For is not saying this simply falling back into the complacencies of organic holism, form mirroring content, 'the perfect match between the problems at stake and the instruments used by Diderot to analyse them', the technologism of genius?[85] It has to be said that Diderot's own attitude to his practice of writing in metaphors does not seem to favour such an interpretation. It is as if, although philosophical discourse cannot construct itself without metaphors, it none the less cancels itself as such by its recourse to them. In the *Lettre sur les Sourds et Muets*, he cuts short the manifold and surprising ramifications of the image of the clock-man (which extend from the laws of harmony to the laws of friendship among the ancient Scythians) by the peremptory sentence:

But I shall abandon this figurative language, which I should only ever use to amuse a child so as to keep its attention from wandering, and shall return to the style [*ton*] of philosophy, *which deals in reasons not comparisons*.[86]

Doubtless, this is a piece of sarcasm aimed at Batteux, as Norman Rudich observes;[87] but the stakes are higher than this. The proverbial expression emphasised by Diderot is taken up by Julie de l'Espinasse in the *Rêve*: 'Listen, doctor, I'll explain what I mean by a comparison; comparisons are almost the only reasons that women and poets can offer'.[88] Figurative speech is thus marginalised as 'feminine', 'poetic' (therefore unscientific); unscientific in itself, at any rate, but admissible as scientific by Bordeu, the man of science:

'Your idea is pefectly correct; but don't you see that it comes to virtually the same thing as that swarm of bees?' To which L'Espinasse replies: 'Why, yes, that's true: I've been talking prose without knowing it'.[89] A philosophical Monsieur Jourdain, she is also like the blind man of Le Puiseaux or Saunderson, speaking more truly than they can, of themselves, know. Is she then, as a woman, another of those Diderotian monsters whose discourse calls into question the security of the 'normal' subject? Or is Bordeu there to stabilise that subject? Is the *Rêve* then a materialist mystical marriage of male science and female poetry – according to the rules of an ideologically suspect canon law of gender?

There is that, I think, in the *Rêve*, although there is not space to go into these issues properly here; just as, despite the persuasiveness of Élisabeth de Fontenay's insistence that the Diderotian body is a federation of autonomous organs governed by a general will,[90] there are also centralising and despotic (albeit enlightened despotic) tendencies in Bordeu's discourse of body politics, with the emphasis on the necessary pre-eminence of the 'point of origin of the cluster', the central nervous system; a pre-eminence with which, moreover, Bordeu himself is identified (identifies himself).[91]

But even if Bordeu's position is in some sense dominant, and if to that extent figurative language is constituted as secondary, ancillary, it is still true that no position in the *Rêve* is completely stable with respect to it. Bordeu himself, as is argued above, is subject to metaphor, and his own discourse allows itself to flirt with figurative speech: he answers d'Alembert's question as to the origin of fictional or poetic elements in narrative with a comparison, itself a response to one of d'Alembert's: 'If you have taken the liberty of comparing the animal to a harpsichord, you will, I am sure, allow me to compare poetic narrative to song', and the fictional elements within it to the harmonics produced by the different notes of the melody.[92] And, as if to sabotage any ready-made equation between gender and the use of figurative language, the explicit eviction of the figural is here left to Julie de l'Espinasse:

But why confuse the question with all this figurative speech? I would say that we each have a different pair of eyes, and therefore see things differently and give different accounts of what we have seen. I would say that every idea triggers off other ideas, and that, depending on one's way of looking at things or on one's character, one either sticks to the ideas that give a plain representation of the fact or brings in the further ideas that they have triggered off; I would say that one can make a selection from these ideas; I would say . . . that it would take a whole book to get to the bottom of this subject alone.[93]

But what this passage also shows is the breakdown of a would-be literal discourse about the mind. Lacking the immediate power of figurative speech to generate ideas, literal discourse has to defer its realisation to writing, to the book of philosophy. But when, in the *Lettre sur les Sourds et Muets*, Diderot seeks to expunge the figural from the text of philosophy, presented instead as a sequence of reasons, he none the less speaks of these reasons as giving rise to a tone, as if reasons are philosophy's tropes.

The full titles of the two *Lettres* mark them out as being for the use of those who can see, those who can hear and speak. How can these texts, and the *Rêve*, keep their promise to the contemporary materialist? Or can he or she no longer see, hear, speak in that way?

If Diderot, for Lenin, anticipates the standpoint of contemporary materialism, this is not in his adoption of specific materialist theses, but in his putting into practice an anti-idealist discursive strategy, one that displaces the syllogism and theoretical argument. The point is even more valid for the materialism(s) – marxist and/or feminist – of today than it was in Lenin's time. Insofar as the *Rêve* appears to authorise particular materialist positions detachable from their textual embodiment, these must be called into question. For Diderot's materialist theses ignore the role of culture in constituting the human subject: 'organisation' is for him a physical rather than a societal concept.

If Diderot is still worth reading for materialism, it can only be because of the textual embodiment of these (questionable) theses. The instances of metaphor are paradigmatic here. For the metaphor refuses to absent itself, to withdraw its own heterogeneity, from the object it constructs. Indeed, metaphor does not even construct an object. Man is not a sentient harpsichord with a spider running about inside. The metaphors construct the human being only in a series of heterogeneous relations with the material world. They thus write off the Cartesian subject, the thinking thing whose relation to bodies, including its own, is not essential to it. The non-fixity of the human in discourse, its non-existence outside the affinities and differences generated by metaphor corresponds to a precariousness of the human species in the cosmos, as a temporary, localised, and contingent configuration of matter, no more and no less monstrous than anything else in nature. Monstrosity is even within the human in the form of the sexual difference – pure difference, not the difference of one sex from the original form of the other, since if 'man is perhaps nothing but an abnormal form of woman, or woman an abnormal form of man', the priority remains uncertain. The sexes do not stand in some teleological relation of complementarity, assuring procreation according to the plan of a Creator.

This is not of course to deny that the discourse of science can and does generate sexual essentialisms distinct from, but in their concrete ideological effects, reinforcing those of religion. Diderot's discourses of gender need close examination, and, as I have said, his science is necessarily blind to the constitutive power of culture, class, gender. Yet his writing (and not only his theses) does more than subvert the dominant philosophical ideology of his day, the 'metaphysico-theological claptrap', as he calls it in the *Rêve*. The displacements of metaphor within his texts, breaking up the unitary subject of humanism, in some sense prefigure and can arguably still contribute to the enterprise of contemporary materialism.

Notes

1 Fontenelle, 'Digression sur les anciens et les modernes', in *Oeuvres*, nouvelle edn., 10 vols., Paris, 1758, IV, p. 170. Translations are all my own.

2 *Ibid.*, p. 171.

3 *Ibid.*, p. 176.

4 *Ibid.*, p. 176.

5 V. I. Lenin, *Materialism and Empirio-Criticism*, Peking (Beijing), 1976, p. 26.

6 A task which Terry Eagleton has engaged with lately, in the chapter 'A small history of rhetoric', in his *Walter Benjamin: Towards a Revolutionary Criticism*, London, 1981, and in his *Literary Theory: an Introduction*, Oxford, 1983, pp. 205–13.

7 For the objections of Hobbes and Gassendi (the *Third* and *Fifth* sets respectively), see Descartes, *Oeuvres Philosophiques*, ed. Ferdinand Alquié, 3 vols., Paris, 1963–73, II, pp. 600–602, 709–11, 775–82. Descartes' original position is set out in the *Sixth Meditation* (*Ibid.*, pp. 487–8); he replies to Hobbes on pp. 602–6, and to Gassendi on pp. 793–4, 833–4.

8 Pascal, *Pensées*, 111 (Lafuma number)/339 (Brunschvicg number). Further references are to the Lafuma number of the fragment in question.

9 *Ibid.*, 115.

10 Spinoza, *Ethics*, Part II, Prop. VII, Note, in *On the Improvement of*

the Understanding; The Ethics; Correspondence, tr. R. H. M. Elwes, New York, n.d., p. 86.

11 Spinoza, *Improvement of the Understanding*, p. 21.

12 *Ibid.*, p.33.

13 J. Locke, *An Essay Concerning Human Understanding*, ed. Peter H. Nidditch, Oxford, 1975, p. 542.

14 *Ibid.*, p. 624.

15 Diderot, *Éléments de Physiologie*, ed. Jean Mayer, Paris, 1964, p. 320.

16 Descartes, *Oeuvres Philosophiques*, II, pp. 787, 790 and 793.

17 *Ibid.*, pp. 775–7.

18 *Ibid.*, p. 417.

19 *Ibid.*, pp. 833–4.

20 Diderot, *Lettre sur les Aveugles*, in *Oeuvres Philosophiques*, ed. Paul Vernière, Paris, 1964, pp. 92, 90.

21 *Ibid.*, p. 82. My emphasis.

22 *Ibid.*, p. 139.

23 *Ibid.*, p. 115.

24 *Ibid.*, p. 83.

25 *Ibid.*, p. 156.

26 *Ibid.*, p. 84.

27 *Ibid.*, p. 85.

28 *Ibid.*, p. 117.

29 *Ibid.*, p. 87.

30 *Ibid.*, p. 87. My emphasis.

31 Descartes, *Oeuvres Philosophiques*, I, p. 654.

32 *Ibid.*, pp. 653–4.

33 Diderot, *Entretiens sur le Fils Naturel*, in *Oeuvres Esthétiques*, ed. Paul Vernière, Paris, 1965, p. 167; *Oeuvres Philosophiques*, pp. 320–21.

34 In *Republic*, 507–509c, Plato compares the sun to the Form of the Good, thus equating the mind with the eye, and sight with knowledge. Sight is picked out as a special sense in that, besides its object (colour), it needs a further element (light) in order to operate (507d). In *Timaeus*, 47, sight is presented by Plato as a divine gift, the precondition of all philosophy, which derives in the first place from the spectacle of regularity in nature, particularly in the heavens. Saunderson is, as it were, responding to this position in accordance with his own, when he says 'If you want me to believe in God, you will have to make me touch him' (Diderot, *Oeuvres Philosophiques*, p. 119). The Platonist valuation of sight above the other senses recurs in Renaissance thought: for Leonardo, the eye is 'the chief and leader' of all the other senses, and 'the window of the soul . . . the chief organ whereby the understanding can have the most complete and magnificent view of the infinite works of nature' (*The Notebooks of Leonardo da Vinci*, ed. Irma A. Richter, Oxford, 1980, pp. 108–10). For Descartes, sight is 'the most

universal and most noble' of all the senses (*La Dioptrique*, in *Oeuvres Philosophiques*, I, p. 651), and 'the most subtle of all the senses', in that it is not set in motion by the air or by other earthly bodies, but by particles of the 'second element', of which the heavens are composed; whereas touch, for the converse reason, is described as the coarsest of all the senses (*Principes de la Philosophie*, IV, 195, 192, in *Oeuvres Philosophiques*, III, pp. 509, 508). And Derrida has shown that Descartes tends to guarantee metaphysical principles by appealing to an unquestioned 'lumière naturelle' ('La mythologie blanche', in Derrida, *Marges de la Philosophie*, Paris, 1972, pp. 318–19); for illustrations see, e.g., Descartes, *Oeuvres Philosophiques*, II, pp. 436, 438; III, p. 102.

35 Diderot, *Oeuvres Philosophiques*, p. 98.

36 *Ibid.*, p. 110.

37 *Ibid.*, p. 111.

38 Diderot, *Oeuvres Philosophiques*, pp. 356–8 and *Oeuvres Esthétiques*, pp. 310–13.

39 Diderot, *Lettre sur les Sourds et Muets*, ed. Paul Hugo Meyer, *Diderot Studies*, VII, 1965, p. 49.

40 *Ibid.*, pp. 50–51.

41 Diderot, *Premières Oeuvres*, eds., Norman Rudich and Jean Varloot, Paris, 1972, Vol. 2, p. 107.

42 Diderot, *Oeuvres Esthétiques*, pp. 164, 167.

43 Diderot, *Lettre sur les Sourds et Muets*, p. 51.

44 Diderot, *Oeuvres Philosophiques*, p. 97.

45 *Ibid.*, p. 85; *Lettre sur les Sourds et Muets*, p. 51.

46 For once I write 'his' rather than 'his or her' because the subject of Western metaphysical reflexion has traditionally been conceived, consciously or unconsciously, as male: to put the matter another way, the characteristic tendency of Western metaphysics to essentialise and also hierarchise difference has contributed to the maintenance of male superiority. Compare Élisabeth de Fontenay's statement that 'the great figures that appear in Diderot's writings – Rameau's Nephew, the Nun, the people born blind, the mathematician in delirium, the savages, the women – undermine the ambition of the Western masculine subject to erect himself into the support of a neutral knowledge and a sovereign power', *Diderot ou le Matérialisme Enchanté*, Paris, 1981, p. 14.

47 In Voltaire's *Micromégas*, the eponymous protagonist has nearly a thousand senses, while his Saturnian travelling-companion has to make do with seventy-two. Yet these seemingly afford them no metaphysical or epistemological insight denied to mere five-sensed earthlings; in fact, they turn out to be Lockeans (Voltaire, *Contes*, eds. Frédéric Deloffre and Jacques van den Heuvel, Paris, 1979, p. 36). *Micromégas* apparently foregrounds difference (creatures a thousand fathoms tall from Saturn, and eight leagues tall from Sirius, the former living 15,000 years and the latter 700

times as long (pp. 19, 21, 23), but neutralises it under a cosmic similitude ('I see differences everywhere', says Micromégas, 'but also proportion everywhere' (p. 24)) which is guaranteed by Providence. The cosmic exoticism merely serves to naturalise the cross-Channel exoticism of English deism, part of the very thought system that Diderot's monsters subvert.

48 Diderot, *Oeuvres Philosophiques*, p. 277.

49 Diderot, *Lettre sur les Sourds et Muets*, pp. 57–62.

50. For uses of the clock-image in Descartes, see *Discours de la Méthode*, Cinquième Partie (*Oeuvres Philosophiques*, I, pp. 623, 631); the letter to the Marquis of Newcastle of 23 November, 1646 (*Ibid.*, III, p. 695); (*Ibid.*, III, p. 955); and in La Mettrie, see *L'Homme Machine*, ed. Paul-Laurent Assoun, Paris, 1981, p. 142.

51 Diderot, *Lettre sur les Sourds et Muets*, p. 62.

52 *Ibid.*, p. 63.

53 Herbert Dieckmann, 'The metaphoric structure of the *Rêve de D'Alembert*', *Diderot Studies*, XVII, 1975, pp. 15–24. Quotation appears on p. 21.

54 *Ibid.*, pp. 68–9 (the discussion of Dieckmann's paper). In his reply, Dieckmann rightly criticises Pappa's trigger-happy use of the word 'poetic' (p. 72), though his own references on p. 71 to Mallarmé and Spender, albeit more carefully phrased, fall arguably into the same essentialising trap. They posit linguistic norms ('the standard prose of philosophic and scientific works of (Diderot's) period'; 'standard' poetry) that can hardly be taken for granted. As Geoffrey Cantor shows, for instance, elsewhere in this volume, the most 'standard' scientific works of the eighteenth century are no less governed by a metaphorical economy than Diderot's.

55 Diderot, *Oeuvres Philosophiques*, pp. 196–8.

56 *Ibid.*, p. 124.

57 See Baillet's account of Descartes' dreams and their interpretation in *Oeuvres Philosophiques*, I, pp. 52–61.

58 The actual expressions 'rêve en montant' and 'rêve en descendant' are in fact used by d'Alembert himself, glossing Bordeu's explanation (*Oeuvres Philosophiques*, p. 360). See also Diderot, *Eléments*, pp. 260, 253.

59 Diderot, *Oeuvres Philosophiques*, p. 361.

60 *Ibid.*, pp. 299–301.

61 Diderot, *Correspondance*, ed. Georges Roth, 16 vols, Paris, 1955–70, III, 1957, pp. 172–3.

62 Diderot, *Oeuvres Philosophiques*, p. 360.

63 *Ibid.*, pp. 328, 297–8.

64 See p. 35, n. 3 of Jean Varloot's admirable edition of the *Rêve de d'Alembert*, Paris, 1971, from which I have derived a good deal more than merely scabrous detail.

65 *Ibid.*, pp. 293–4.

66 *Ibid.*, pp. 271–2. This image also echoes La Mettrie, although it is

used for a different purpose: 'Just as a violin string or the key of a harpsichord vibrates and gives out a sound, the strings of the brain, on being struck by sound waves, were stimulated into uttering or repeating the words that affected them' (*L'Homme Machine*, p. 111). Two points about the translation: it may seem odd to speak of the 'key' of a harpsichord as vibrating; what vibrates is of course the string plucked by the quill. However, La Mettrie's word is *touche*, which in the case of keyboard instruments can only mean '(finger-) key', and I have stuck to a literal translation. The past tense 'were' is explained by the fact that La Mettrie is engaged in the familiary eighteenth-century exercise of attempting to reconstruct the origins of a phenomenon, in this case language.

67 *Ibid.*, p. 272.

68 *Ibid.*, p. 273.

69 Diderot, *Éléments*, p. 305.

70 Diderot, *Lettre sur les Sourds et Muets*, p. 62.

71 Diderot: *Oeuvres Philosophiques*, p. 273; *Éléments*, p. 243.

72 Diderot, *Oeuvres Philosophiques*, p. 321.

73 *Ibid.*, p. 293.

74 *Ibid.*, pp. 296, 298, 328.

75 *Ibid.*, pp. 298, 272.

76 Lenin, *Materialism*, pp. 311 (his emphasis), 312.

77 *Ibid.*, pp. 162–3.

78 Diderot, *Oeuvres Philosophiques*, p. 564.

79 Diderot denies the homogeneity of matter in *De l'Interprétation de la Nature*, *Oeuvres Philosophiques*, p. 239 and the *Principes Philosophiques sur la Matière et le Mouvement*, *ibid.*, p. 394.

80 *Ibid.*, p. 306.

81 According to Bordeu, this centre is housed in the meninges, the membranes enveloping the brain and spinal cord, one of which is actually called the arachnoid, from the Greek for spider, or alternatively cobweb: it may be then that this particular metaphor was generated by etymology, the signifier taking precedence over a conscious desire to express the nature of reality.

82 *Ibid.*, p. 314.

83 Derrida, *Marges*, pp. 254–5.

84 De Fontenay, *Diderot*, p. 155.

85 Gabrijela Vidan, 'Diderot: la construction scientifique et son relais par l'imagination', *Studies on Voltaire and the Eighteenth Century*, 155, 1976, p. 2222.

86 Diderot, *Lettre sur les Sourds et Muets*, p. 63. Diderot's emphasis.

87 Diderot, *Premières Oeuvres*, II, p. 124, n. 1.

88 Diderot, *Oeuvres Philosophiques*, p. 307. The allusion, as in the last quotation, is of course to the French proverb 'Comparaison n'est pas raison'.

89 *Ibid.*, p. 314.

90 De Fontenay, *Diderot*, pp. 156–7.
91 Diderot, *Oeuvres Philosophiques*, p. 359.
92 *Ibid.*, p. 368.
93 *Ibid.*, p. 368.

9 *Stephen Copley*

Polite culture in commercial society

In the course of the *Essays Moral Political and Literary*, David Hume provides a series of elaborate paradigmatic descriptions of the operation and development of commerce in society, and of its political and cultural consequences. In 'Of Commerce' and 'Of Refinement in the Arts' in particular, he represents the progress of commercial society as a process of interrelated economic expansion, political development and cultural refinement. His formulations had a considerable influence on the development of later eighteenth-century Scottish academic study in many of the new and increasingly distinctive fields of economic, political and cultural inquiry that emerged in the period. As J. G. A. Pocock has written, 'the great Scottish school of social philosophers, who are in a complex fashion his [Hume's] immediate heirs . . . were greatly aided by his teachings'.[1]

In this essay I want to examine a number of interesting features in the ideological construction of the paradigms provided by Hume, looking in particular at the place he accords to refinement in the arts in society as a whole. Some general points are worth making to begin with. E. A. J. Johnson, writing in the 1930s, points out that, in terms of economic theory, Hume's *Essays* represent a fusion of a considerable amount of 'summary' and 'synthesis' of 'the English economic thought of his time', and a degree of striking innovation; he warns that, 'there is a stylistic quality of Hume's writings which leads one to believe that he is stating essentially new ideas, although, at closer inspection, many of these apparent innovations are subtle, compact, and felicitous restatements of older views'.[2] In a more recent commentary, E. J. Hundert considers one of these areas of real or apparent innovation, which has a particular relevance to this essay. In a short journal article he continues an earlier contributor's discussion of the importance of Hume's 'analysis of psychological motivation' in his account of the economy, particularly as it affects his treatment of 'working men'.[3] He acknowledges that this analysis can be somewhat contradictory, writing that, 'although his analysis of the worker and his motivation ran counter to current opinion, some of Hume's most fundamental assumptions about the working man were identical to those of his contemporaries'; and cites his agreement with Sir William Temple 'that the behaviour of workers significantly differed from that of other men' in support of this. Nevertheless, he claims that

> The radical nature of Hume's analysis lay in its psychological egalitarianism. Hume made no distinction between the natural inheritance of different classes of men. He held that all men first and foremost seek happiness, and that this primarily consisted of action and pleasure. In order to achieve happiness men worked, for in working toward a certain material reward, pleasure, action, and avarice were satisfied, while difficulty was overcome. In the world of work and achievement men confronted their basic desires and thus gained the opportunity of satisfying them. Indeed, work is initially undertaken to satisfy the passions common to all men. In this esentially natural and sensuous process men learn to enjoy the activity itself.[4]

This certainly identifies an important strand of argument in the *Essays*. All through, Hume insistently traces economic behaviour back to its psychological foundations, and generalises about human nature and motivation in terms which cross any class divides in society. In this he departs not only from writers such as Temple, but also from Mandeville, whose analysis of luxury in other respects forms an important – if severely qualified – starting point for his economic arguments.[5] Both these writers share the widespread assumptions of economic writers of the time about the necessity of maintaining a hierarchic division in society, between citizens and consumers on the one hand, and subsistence labourers who are driven only by their animal needs on the other; and both insist on the need for an approach to social and economic policy that is sharply differentiated in its treatment of these two classes. For Hume, in contrast, the premise of general psychological egality in society is matched in a number of the essays by advocacy of a degree of economic egalitarianism that is innovative in the period – as when he writes in 'Of Commerce' that 'every person, if possible, ought to enjoy the fruits of his labour, in a full possession of all the necessaries, and many of the conveniences of life. No one can doubt but such an equality is most suitable to human nature, and diminishes much less from the happiness of the rich, than it adds to that of the poor'.[6]

In spite of the impression made by passages such as this, however, we must recognise that the extent of Hume's advocacy of social egalitarianism in the *Essays* as a whole is severely limited, as is his commitment to the 'psychological egalitarianism' that Hundert claims underlies it. Both are in many ways more striking in their appearance than they are in their substance. Despite the generalising and egalitarian rhetoric that pervades a number of the essays, the discussions that are conducted in them are shot through with implicit admissions of distinction and division in society – admissions which are often integral to Hume's arguments, but which potentially threaten to undermine the generalisations on which those arguments depend. Hundert's summary fails to take full account of the extent to which these admissions qualify and condition our reading of the essays. John Robertson provides a more complex description of the arguments that are conducted in them.[7] He acknowledges the new directions in which these arguments are apparently leading, while at the same time drawing attention to some of the areas in which Hume's accommodations of social division might seem to conflict with them. He differentiates Hume from the earlier writers mentioned above, and writes that 'what Hume appears to suggest is that within the framework of continuing social division, commerce will in the long run make it possible for every individual to satisfy the material and moral requirements of citizenship'. However he notes that these suggestions remain only as unrealised 'projections into the future', and that they are 'not elaborated in any detail, but are left rather as implications than as explicitly stated conclusions of the analysis of the social and moral dimensions of progress'; and he remarks finally that Hume does not 'reach the point of asking' whether commerce itself might not create 'new forms of degradation and corruption' in society as it develops.[8]

The points outlined here serve to draw our attention to a number of important general features of Hume's style of writing in the *Essays*, which we must take into account in our reading of the arguments that he puts forward in them. If we examine these arguments closely, we find that they often incorporate disjunctures such as the one Robertson points out, between passages of analysis, and summaries of the implications that are derived from that analysis; that Hume's conclusions often draw as strongly on those implications as on the substantial proofs that he provides; and that in his discussions he repeatedly avoids considering a range of questions that might otherwise seem central to the matters he raises. These and other stylistic features of the *Essays* have tended to be ignored by commentators. We can best examine the consequences that they have in Hume's arguments, first, by looking in detail at a number of passages from the essays themselves; and second, by comparing Hume's discussion of society, the economy and 'the arts' with a later analysis by Ferguson, which owes a considerable amount to him, but in which a range of divisions and distinctions that are occluded in the *Essays* are made explicit, and are subjected to critical analysis.

In approaching the *Essays* in this way, I want to take two things in particular into account. The first is a consideration that we must bear in mind when we approach any eighteenth-century social-philosophical or economic writings. J. G. A. Pocock has demonstrated the importance of reading these writings in the light of the preoccupations that inform a tradition of civic humanism, which has its roots in Renaissance readings of classical republican texts. However he has also shown that the eighteenth-century works reveal every sign of being constructed in a period of marked discursive transition, in which the political, economic and moral discourses available in that earlier tradition are gradually displaced by or transformed into the new discourses of political economy. As he writes, 'the rapidly developing style of political economy . . . is the dominant mode of Augustan political thought; but its preoccupation with 'virtue' as well as with economics 'is so great that we have to recognize that the first chapter in the history of political economy is also a further chapter in the history of civic humanism'.[9]

In the case of the *Essays*, the legacy of civic humanism is strongly evident. In many of them, political progress is analysed at least in part in relation to the achievement of liberty in the state, the defence of it against corruption, and the cultivation of civic and martial virtues in the population; and 'the arts' are examined and assessed for their contribution to those ends. At the same time, however, Hume's arguments depart from that tradition in several striking ways. Pocock mentions the extent to which their commitment to 'progress' minimises their concern with 'corruption', and so re-casts a central dialectic of humanist thought.[10] Robertson traces an emphasis on the individual in them, which has its roots in a tradition of Scottish jurisprudential writing, but which, taken along with the new stress on progress, marks a considerable step in the emergence of liberal from humanist political discourse.[11] Finally the *Essays* provide an influential stimulus in the progressive redefinition of terms that marks the emergence of political economy from civic humanism. They must therefore be read with some acknowledgement of the potentially changing and changeable status of some of the key terms they employ. These terms – luxury, liberty, arts – can have crucially different associations in the different discourses in which they are deployed. At various points, Hume's arguments depend on his exploitation of the ambiguities of meaning that this gives rise to, either because of his implicit refusal to acknowledge important distinctions between the terms he uses, or because of his effective and unacknowledged elisions of different associations of meaning within each term. Analysis of these ambiguities will give us one means of coming to terms with the ideological implications of the account of society that he provides.

One of the terms mentioned above that recurs in various of the *Essays* and that is particularly open to critical scrutiny, is 'art'. A brief discussion of this term will serve to introduce the second set of considerations that I want to bring to bear in my reading of Hume's account of society. In the study referred to earlier, E. A. J. Johnson notes the frequency with which the term 'art' and its cognates are employed in seventeenth- and eighteenth-century English economic writings, and catalogues some of its available meanings.[12] He writes that in one group of these meanings, 'art' originally signifies general skill in workmanship, and that it only gradually comes to acquire its more restricted modern association with skill in aesthetic works and subjects of taste; in another related sense its meaning is similarly 'both expanded and restricted' from its original suggestion of 'anything wherein skill may be attained or displayed' to later particularised applications such as 'black art', 'mechanical art' or 'the art of conversation'.[13] We can add various other specific ramifications of the word in its eighteenth-century usage, where it can cover the 'higher arts' of government, statecraft and soldiership; the 'liberal arts'; the 'mechanical arts' of agriculture and manufacture, and the 'polite arts' of conversation and taste.

The important thing to notice about this range of meanings, as they appear in the *Essays*, is that they can be seen to describe activities in any one of a number of distinct categories in the economy, and so can bear any one of a number of relations to labour, luxury, leisure, production or consumption in society. The phrase 'the practice of arts' can mean in some circumstances 'the practice of skilled labour' – and so describe an aspect of production. At the same time, if it is used to describe 'the practice of the polite arts', it can refer to the occupation of leisure, and so – when it includes the 'arts' of 'taste' and 'appreciation' – define an aspect of consumption. As we will see, Hume draws on all these available meanings and associations of the word in the course of his arguments, at times eliding them and at times distinguishing between them. In particular, in several of the essays he relies on treating 'the arts' as a single category, while at the same time maintaining important distinctions between the 'mechanical arts' on the one hand and the 'liberal' and 'polite' arts on the other. These distinctions appear most interestingly in his treatment of the 'polite' arts, and it is in this area that I want to draw attention to the second group of considerations that we must bear in mind when reading the *Essays*.

Nicholas Phillipson and others have drawn attention to the importance that ought to be accorded in studies of the Scottish Enlightenment to the emergence of a middle-class 'polite' culture in Scotland after the Union of 1707; and to the enormous influence that English 'polite' periodical literature – and in particular Addison and Steele's *The Spectator* – had on that development.[14] Taking up these points, Pocock comments on the 'massive and rapid adoption of an Addisonian Whig political, or rather social, culture' in Scotland in the period, and cites in confirmation the proliferation of 'Spectatorial clubs and societies' in Edinburgh, dedicated to 'practising the virtues of polite conversation and enlightened taste while discussing the economic, cultural and . . . moral improvement of Scottish life'.[15] Phillipson notes that Hume's *Essays* show every sign of being part of this movement, and that they are modelled very self-consciously on the earlier periodical essays. He writes of them that although 'Hume once called Addison's essays 'triffling' . . . he owed more to Addison than historians have been accustomed to admit'; and he continues by suggesting that in Hume's hands 'the polite Addisonian essay' develops 'as a sophisticated and flexible mode of moral discourse, capable of attracting an intelligent salon and coffee-house readership as well as philosophers and men of letters'.[16]

The *Essays* are marked in a number of interesting ways by the circumstances of their production. At the most straightforward level, a considerable amount of space in them is devoted to a thoroughly partisan discussion of the nature and place in society of the bourgeois polite culture of which they are a part. Beyond that, they reveal a whole range of the characteristic postures and preoccupations of earlier periodical writing. Beyond that again, at various stages Hume exploits the form and mode of address of the polite essay itself, in order to condition and qualify the arguments that he puts forward in them. In particular, he in large part adopts the stance of the 'polite' essayist from earlier periodical essays; defines a constituency of 'polite' readers for whom he is writing; and silently incorporates a range of assumptions about the nature, capacities and status of those readers, and about his own privileged relation to them, into his discussions of society.

The debt owed by the *Essays* to the earlier periodicals is acknowledged in the Advertisement to the first edition, where Hume writes:

> MOST of these ESSAYS were wrote with a View of being publish'd as WEEKLY-PAPERS, and were intended to comprehend the Designs both of the SPECTATORS and CRAFTSMEN. But having dropt that Undertaking, partly from LAZINESS, partly from WANT OF LEISURE, and being willing to make Trial of my Talents for Writing, before I ventur'd upon any more serious Compositions, I was induced to communicate these Trifles to the Judgement of the Public. . . .[17]

He goes on to describe their form in a way which draws our attention to some of the characteristics of the periodical literature to which they claim to owe a debt:

> THE READER must not look for any Connexion among these ESSAYS, but must consider each of them as a Work apart. This is an Indulgence that is given to all ESSAY-WRITERS; and, perhaps, such a desultory Method of Writing, is an equal Ease both to AUTHOR and READER, by freeing them from any tiresome Stretch of Attention and Application.[18]

In line with the model of the periodical, the Advertisement offers to provide the polite reader with approachable, self-contained, non-specialised accounts of areas of non-vocational knowledge, presented as improving or educative entertainment that it will not require 'any tiresome Stretch of Attention and Application' to master. Incidentally, it also provides a framework that accommodates the shifts and ambiguities of terminology that run through the Essays, by instructing 'THE READER' not to look for 'any Connexion' between them – an injunction that we will later violate by examining some of the terms that are employed throughout, as they appear in a number of the different essays. For the time being, however, we can find a focus for examining Hume's construction of a 'polite' perspective on society by looking first at some passages in his acknowledged model, *The Spectator*, in which various matters germane to his arguments are raised.

The first issue of *The Spectator* opens with an introduction to the collective authorial persona of 'Mr Spectator'. He writes, 'I have observed, that a Reader seldom peruses a Book with Pleasure, 'till he knows whether the Writer of it be a black or fair Man, of a mild or cholerick Disposition, Married or a Batchelor, with other Particulars of the like nature, that conduce very much to the right Understanding of an Author'. He does not provide us with quite these details, however. Instead he gives an account of his social identity, which is guaranteed by his ownership of land ('I was born to a small Hereditary Estate, which, according to the Tradition of the Village where it lies, was bounded by the same Hedges and Ditches in William the Conqueror's Time that it is at present . . .'); and then describes his salient characteristic of taciturnity, his 'insatiable Thirst after Knowledge', and his acquaintance with various aspects of City life. These all contribute to his description of his position as essayist:

> Thus I live in the World, rather as a Spectator of Mankind, than as one of the Species; by which means I have made myself a Speculative Statesman, Soldier, Merchant and Artizan, without ever medling with any Practical Part in Life. I am very well versed in the Theory of an Husband, or a Father, and can discern the Errors in the Oeconomy, Business and Diversion of others, better than those who are engaged in them; as Standers-by discover Blots, which are apt to escape those who are in the Game. . . . In short, I have acted in all the Parts of my Life as a Looker-on, which is the character I intend to preserve in this Paper.[19]

This reveals two particularly interesting aspects of the essayist's persona. First, it shows his secure social position and confidence in his identity of interests and perspective with his readers; and secondly, it indicates his position of informed awareness of all the 'practical parts' of life, but his privileged exclusion from active participation in them, which gives him a special status as a spectator on society. Both are used to define the peculiar and privileged relation between 'polite' essayist and 'polite' reader. In part, this definition is a matter of admitting the social exclusivity of the perspectives on society offered in 'polite' writing, and the similar exclusivity of the audience for which it is designed. Despite the essayist's suggestion that he has access to a multiplicity of social perspectives, those that he offers are far more limited than he originally announces; they never, for example, stretch as far as to include those of 'the Artizan', who is mentioned above. Similarly, although he suggests in his description of the members of the Spectator Club in the second and third issues of the periodical that his catchment of readers represents a universal cross-section of society, the description itself simultaneously defines the exclusive bounds of the classes from which 'the polite' are constituted.[20]

More interestingly from our point of view, the 'polite' readers of the periodical are repeatedly treated as a special category in society, like the essayist abstracted by virtue of their 'politeness' from involvement in the 'practical parts of life'. In No. 509 a letter on money from 'Hezekiah Thrift' is introduced with a word of apology:

> The useful Knowledge in the following Letter shall have a Place in my Paper, tho' there is Nothing in it which immediately regards the Polite or the Learned World; I say Immediately, for upon Reflection every Man will find there is a remote Influence upon his own Affairs, in the Prosperity or Decay of the Trading Part of Mankind.[21]

The readers are implicitly congratulated as belonging to 'the Polite or the Learned World', which is set aside from concern with the practical business of the economy, as it is carried out by 'the Trading Part of Mankind' – even though at other times the readership is assumed to include members of that 'Trading Part' of society.

A number of other issues discuss the special position of polite culture in society and enumerate the advantages of cultivating polite learning and sensibility. No. 287 begins with an enconium on the British constitution as the source of generally diffused liberty in society. The argument appears egalitarian in its insistence on the common humanity – and consequent common political rights – of the subjects of the state:

> That form of Government appears to me the most reasonable, which is most conformable to the Equality that we find in Human Nature, provided it be consistent with Publick Peace and Tranquillity. This is what may properly be called Liberty, which exempts one Man from Subjection to another, so far as the Order and Oeconomy of Government will permit. . . . Liberty should reach every Individual of a People, as they all share one common Nature. . . .[22]

However, as the essay proceeds to a discussion of 'Riches and Plenty' and 'the liberal Arts', it admits of economic distinctions in society which go against that egalitarian rhetoric. Interestingly these centre on an admission that the economic foundation of polite culture must be a leisure that is based on the labour of others:

The first thing every one looks after, is to provide himself with Necessaries. This Point will engross our Thoughts till it be satisfied: If this is taken Care of to our Hands, we look out for Pleasures and Amusements; and among a great number of idle People, there will be many whose Pleasures will lie in Reading and Contemplation. These are the two great Sources of Knowledge, and as Men grow wise, they naturally Love to communicate their Discoveries; and others seeing the Happiness of such a Learned Life, and improving by their Conversation, emulate, imitate and surpass one another, till a Nation is filled with Races of wise and understanding Persons. Ease and Plenty are therefore the great Cherishers of Knowledge. . . .[23]

Knowledge of the polite and liberal arts is one of the 'Pleasures' which are a product of leisure, and which are supported by labour elsewhere in society; but in its effects the acquisition of that knowledge differentiates those involved from the purely 'idle'. The point is made particularly tellingly in No. 411. 'Mr Spectator' catalogues the non-material advantages of a 'Polite Imagination' by analogy with the solidly material delights of property ownership, and distinguishes between those to whom these advantages are open, and the rest of society:

A Man of a Polite Imagination, is let into a great many Pleasures that the Vulgar are not capable of receiving. He can converse with a Picture, and find an agreeable Companion in a Statue. He meets with a secret Refreshment in a Description, and often feels a greater Satisfaction in the Prospect of Fields and Meadows, than another does in the Possession. It gives him indeed a kind of Property in everything he sees, and makes the most rude uncultivated Parts of Nature administer to his Pleasures: So that he looks upon the World, as it were, in another Light, and discovers in it a Multitude of Charms, that conceal themselves from the generality of Mankind.[24]

He then lays out the differences between the idleness that leads to vice and folly, and the innocent pleasure of cultivating the imagination:

There are, indeed, but very few who know how to be idle and innocent, or have a Relish of any Pleasures that are not Criminal; every Diversion they take is at the Expence of some one Virtue or another, and their very first Step out of Business is into Vice or Folly. A Man should endeavour, therefore, to make the Sphere of his innocent Pleasures as wide as possible, that he may retire into them with Safety, and find in them such a Satisfaction as a wise Man would not blush to take. Of this Nature are those of the Imagination, which do not require such a Bent of Thought as is necessary to our more serious Employments, nor, at the same Time, suffer the Mind to sink into that Negligence and Remissness, which are apt to accompany our more sensual Delights, but, like a gentle Exercise to the Faculties, awaken them from Sloth and Idleness, without putting them upon any Labour or Difficulty.[25]

As it is described here, the domain of the polite is marked by the special advantages that are conferred on its members. It is exclusive rather than general, and develops from the enjoyment of leisure rather than from the pursuit of business. It therefore involves the cultivation of pleasure rather than the expenditure of labour. However, the development of polite taste cannot be dismissed as an encouragement to sloth or idleness, because it does involve 'a gentle Exercise to the Faculties' of those involved. Based on that, it can be afforded the special and privileged status that we have indicated.

In Hume's *Essays*, the legacy of assumptions about polite culture that we have outlined above emerges in a variety of ways. We can recognise it most immediately by examining the use in a number of the essays of an interesting form of exclusive address, which affords the polite reader a literally privileged position in the arguments that Hume conducts. We can trace the formation of this address most clearly in the first essay, 'Of the Delicacy of Taste and Passion'. The essay begins with a paragraph of generalisation on the psychological bent of certain people: 'Some people are subject to a certain delicacy of passion, which makes them extremely sensible to all the accidents of life, and gives them a lively joy upon every prosperous event, as well as a piercing grief when they meet with misfortune and adversity'.[26] These hypothetical people are then compared to equally hypothetical 'men of cool and sedate tempers', and Hume concludes that 'I believe, when every thing is balanced, there is no one who would not rather be of the latter character, were he entirely master of his own disposition'.[27] In the second paragraph, the psychological generalisation of the first is extended to cover 'delicacy of taste', which is presented as being analogous to 'delicacy of passion' in various ways: 'There is a delicacy of taste observable in some men, which very much resembles this delicacy of passion, and produces the same sensibility to beauty and deformity of every kind, as that does to prosperity and adversity, obligations and injuries'.[28] Hume then describes the effects of aesthetic experience for 'a man possessed of this talent'.[29] As the paragraph in which he does this develops, it incorporates an interesting change of pronouns, which shifts attention away from the 'man' who is described initially:

> A polite and judicious conversation affords him the highest entertainment; rudeness or impertinency is as great punishment to him. In short, delicacy of taste has the same effect as delicacy of passion. It enlarges the sphere both of our happiness and misery, and makes us sensible to pains as well as pleasures which escape the rest of mankind.[30]

The detached description of characteristics 'observable in some men', with which the paragraph began, has developed into an exclusive definition of the capacities shared by the essayist and the polite readership of the essay, for whom he acts as a spokesman; and has come to suggest that there is an opposition between that readership and 'the rest of mankind', who cannot participate in their 'pains' and 'pleasures'.

The most striking of the distinctions that are drawn between the two groups that are defined here follow on from the claim that 'when a man is possessed of that talent [delicacy of taste], he is more happy by what pleases his taste, than by what gratifies his appetites, and receives more enjoyment from a poem, or a piece of reasoning, than the most expensive luxury can afford'.[31] Hume goes on to enumerate some of the pleasures of such cultivated taste, and then offers to retract an impression that he may have conveyed. In the process he reinforces that impression, while only partially qualifying it in his subsequent argument. He writes:

> But perhaps I have gone too far, in saying that a cultivated taste for the polite arts extinguishes the passions, and renders us indifferent to those objects which are so fondly pursued by the rest of mankind. On further reflection, I find, that it rather improves our sensibility for all the tender and agreeable passions; at the same time that it renders the mind incapable of the rougher and more boisterous emotions.[32]

The rest of the essay then reinforces the distinctions that mark off the person of taste from the rest of mankind. We are told that 'the study of the beauties either of poetry, eloquence, music, or painting' gives 'a certain elegance of sentiment to which the rest of mankind are strangers'; and that it excites emotions which are 'soft and tender' and 'draw off the mind from the hurry of business and interest;

cherish reflection; dispose to tranquillity; and produce an agreeable melancholy, which, of all dispositions of the mind, is the best suited to love and friendship'. Finally we are assured that 'a delicacy of taste is favourable to love and friendship, by confining our choice to few people, and making us indifferent to the company and conversation of the greater part of men'.[33]

All the suggestions in this final part of the essay have interesting ramifications when they are set against the accounts of the working of the economy that Hume provides in various of the later essays. The most important of these suggestions are his repeated claims for the dissociation of aesthetic 'taste' from the 'appetites' and 'passions' – claims that are put forward most emphatically when it is suggested (albeit with a qualifying disclaimer) that the development of aesthetic sensibilities 'extinguishes the passions', and so 'renders us indifferent to those objects which are so fondly pursued by the rest of mankind'. The significance of these claims can be seen if we think for a moment of the importance of the 'appetites' and 'passions' in the later discussions of the working of the economy. As we have already seen, these discussions continually make reference back to Hume's analysis of the psychological foundations of human economic behaviour, which he describes largely in terms of the operation of these 'appetites' and 'passions'. His account of their effects is complex. He repeatedly presents the self-interested 'passions' as the psychological mainsprings of the luxury economy, treating them as universal and unmediated impulses to particular forms of economic behaviour, even while he suggests that other interests and motives are dependent on externally determined social circumstances. In 'Of the Rise and Progress of the Arts and Sciences', for example, he distinguishes between the forms of progress that can be expected in commerce and in learning, and explains the differences in the following terms:

> Avarice, or the desire of gain, is a universal passion, which operates at all times, in all places, and upon all persons: but curiosity, or the love of knowledge, has a very limited influence, and requires youth, leisure, education, genius, and example, to make it govern any person.[34]

At other times he provides a more complex account of the nature and evolution of the self-interested passions, in which they too are acknowledged to be socially conditioned. In 'Of Interest' he in effect traces the process of construction of the 'passion' of avarice, from the basis of a quite different 'desire', suggesting that it affects the merchant in particular, even though he had elsewhere claimed it to be 'universal'. He writes:

> There is no craving or demand of the human mind more constant and insatiable than that for exercise and employment; and this desire seems the foundation of most of our passions and pursuits. Deprive a man of all business and serious occupation, he runs restless from one amusement to another. . . . Give him a more harmless way of employing his mind or body, he is satisfied, and feels no longer that insatiable thirst after pleasure. But if the employment you give him be lucrative, especially if the profit be attached to every particular exertion of industry, he has gain so often in his eye, that he acquires, by degrees, a passion for it, and knows no such pleasure as that of seeing the daily increase of his fortune.[35]

Further implicit admissions that the 'passions' on which the economy depends may be socially determined and differentiated emerge in a number of the other essays on economic issues. We can trace one such admission in a central passage in 'Of Commerce', where it underlies one of Hume's most succinct formulations of the relation between psychology and economic behaviour.[36] The passage follows a similar pattern to that which we have traced in 'Of the Delicacy of Taste and Passion'. It begins with a generalisation. Hume writes that, 'every thing in the world is purchased by labour; and our passions are the only causes of labour', and then goes on to discuss the importance of luxury in stimulating those passions, increasing 'industry', and so ultimately augmenting 'the greatness of the sovereign, and the happiness of the state'.[37] As he does so, he departs from his initial inclusive generalisation about 'our passions', and effectively excludes the reader from the consequences of that generalisation. This emerges first in his discussion of the behaviour of 'labourers', a discussion which is extended to include 'the other members of the state', but which differentiates them all from the unparticularised reader who is addressed as 'you':

> It is a violent method, and in most cases impracticable, to oblige the labourer to toil, in order to raise from the land more than what subsists himself and family. Furnish him with manufactures and commodities, and he will do it of himself; afterwards you will find it easy to seize some part of his superfluous labour, and employ it in the public service, without giving him his wonted return. Being accustomed to industry, he will think this less grievous, than if at once you obliged him to an augmentation of labour without any reward. The case is the same with regard to the other members of the state. . . .[38]

This division becomes more marked in the next paragraph, where the classical humanist virtues of 'martial . . . genius' and 'passion for public good' are discounted as possible informing principles of civic behaviour in modern society, because they are 'too disinterested and too difficult to support'. Instead we are told that 'it is requisite to govern men by other passions, and animate them with a spirit of avarice and industry, art and luxury'.[39] The 'passions' are no longer presented as being universal determinants of action. Instead they are allowed to be socially differentiated, and, in this instance, to be manipulated as a mechanism of social control by those who are presumably not affected by them, in order to 'govern' and 'animate' the rest.

If we return to 'Of the Delicacy of Taste and Passion' in the light of these economic arguments, we can trace the consequences of the patterns of exclusion that we have noted in Hume's writing. Broadly, we can recognise that in that essay – as in various other essays in the rest of the collection – Hume identifies his constituency of readers as members of polite culture, and implicitly and explicitly marks out the domain of that polite culture as an area of privileged exclusion from the demands of the economy, from the generalisations about human nature on which his explanations of the working of that economy depend, and from the psychological compulsions that are instilled within it. We can confirm this by looking again at the divisions that we identified between the readership and 'the rest of mankind' in the last part of the essay.

In the initial distinction between 'taste' and 'appetite', quoted above, 'enjoyment' of 'a poem, or a piece of reasoning' is differentiated from the pleasure afforded by 'expensive luxury'. The cultivation of 'taste' is seen as opening the way to aesthetic pleasures that cannot be categorised under the heading of luxury, as they can be satisfied without being directly linked to increased material consumption. In this respect, the man of developed aesthetic sensibility can be represented as being dissociated from the materiality of the economy in which he lives, which depends for its operation on the existence and indulgence of luxury. Even when Hume qualifies his statement about the development of 'taste' extinguishing the 'passions', the formulation that he substitutes still distinguishes between the effects of the 'tender and agreeable passions' that he now says are stimulated by a 'taste for the polite arts', and those that are excluded, which are exactly the ones that are elsewhere said to be necessary to stimulate the economy. The 'soft and tender' emotions 'excited' by 'the beauties either of poetry, eloquence, music, or painting' encourage 'reflection' and 'tranquillity', and 'draw off the mind from the hurry of business and interest'. The cultivation of 'taste' remains a 'polite' refinement of leisure. In its end result, it not only differentiates 'the polite' from the rest of society, but it justifies their leisure in an economy in which men are elsewhere stimulated by 'interest' to ever-increasing productive 'action'. The connection that Hume does not make is to follow through his statement in 'Of Commerce', and define that leisure as being itself a luxury component in the economy, which, like 'every thing in the world', must necessarily be 'purchased by labour' somewhere in society.

We can follow up some of the general consequences of these implicit divisions and distinctions in Hume's arguments by examining the first part of 'Of Refinement in the Arts', in which he provides one of his fullest endorsements of the progress in human happiness that is made possible by the operation of a luxury economy in society. The essay begins with a discussion of the nature of luxury, which is presented as a problematic term 'which may be taken in a good as well as in a bad sense'. It is defined initially as meaning 'great refinement in the gratification of the senses'[40] – a definition that escapes in large part from the morally condemnatory overtones with which the term is still usually loaded in mid-eighteenth-century usage. This effect is confirmed when Hume goes on to adopt his own defining term – 'refinement' – in his subsequent argument about the beneficial effects of luxury, and draws on the positive connotations of that term to colour his argument throughout. This is immediately apparent as the essay continues with a discussion of 'the effects of refinement both on *private* and on *public* life'.[41] The sequence in which these are considered confirms the shift in emphasis that we have already identified, giving primacy to the private domain of the individual, and only then continuing into the public and civic domains in which the classical discourses of humanism are formulated. Hume's discussion thus begins with an account of the components of 'human happiness', apparently considered in universal psychological terms, and only later goes on to consider 'public' questions such as whether the arts contribute to liberty or corruption in society.

John Barrell has drawn attention to an interesting ambiguity of phrasing in the initial discussion of 'happiness'.[42] Hume writes:

Human happiness, according to the most received notions, seems to consist in three ingredients: action, pleasure, and indolence: and though these ingredients ought to be mixed in different proportions, according to the disposition of the person; yet no one ingredient can be entirely wanting, without destroying, in some measure, the relish of the whole composition.[43]

As Barrell points out, this sentence seems to be concerned primarily with the psychological make-up of the individual, employing 'disposition' in the sense that we have already encountered in 'Of the Delicacy of Taste and Passion'. However, there is a strong secondary suggestion that the whole sentence relates to the placement of figures in a pictorial composition, and is prescriptive rather than descriptive. In this sense the proportionate mixture of 'action, pleasure, and indolence' in the individual 'ought to be' determined by the 'disposition' of the 'person' in the whole design, which is apprehended aesthetically by the observer, for whom the lack of particular 'ingredients' destroys 'in some measure, the relish of the whole composition'. The sentence reproduces and extends the ideological postures that we have described in other passages earlier in this paper. Its explicit insistence on the need for a mixture of 'action, pleasure and indolence' in every individual represents a level of psychological universalisation across the social hierarchy which runs counter to the divisions between, on the one hand, citizens and consumers, and on the other, subsistence producers, that we mentioned as informing much earlier economic writing. At the same time it implicitly marks off the essayist as observer through the motif of his aesthetic perception of the design of society, in terms which develop the patterns of implicit dissociation that we encountered earlier; and in doing so, it accommodates a more rigidly hierarchic account of the social order than the other reading had suggested.

The ensuing passage of the essay is worth quoting at length. In it Hume leads on from the generalisations we have just looked at, and insists that 'action' and 'pleasure' are the dominant sources of satisfaction in human life, while 'indolence' is not in itself enjoyable but, 'like sleep, is requisite as an indulgence, to the weakness of human nature, which cannot support an uninterrupted course of business or pleasure'. He continues:

Education, custom, and example, have a mighty influence in turning the mind to any of these pursuits; and it must be owned that, where they promote a relish for action and pleasure, they are so favourable to human happiness. In times when industry and the arts flourish, men are kept in perpetual occupation, and enjoy, as their reward, the occupation itself, as well as those pleasures which are the fruit of their labour. The mind acquires new vigour; enlarges its powers and faculties; and, by an assiduity in honest industry, both satisfies its natural appetites, and prevents the growth of unnatural ones, which commonly spring up, when nourished by ease and idleness.[44]

He then introduces a distinction between 'mechanical' and 'liberal' arts, but insists on their mutual dependence and simultaneous development in society under the influence of the 'spirit of the age':

Another advantage of industry and of refinements in the mechanical arts, is, that they commonly produce refinements in the liberal; nor can one be carried to perfection, without being accompanied, in some degree, with the other. The same age which produces great philosophers and politicians, renowned generals and poets, usually abounds with skilful weavers, and ship-carpenters. We cannot reasonably expect, that a piece of woollen cloth will be wrought to perfection in a nation which is ignorant of astronomy, or where ethics are neglected. The spirit of the age affects all the arts, and the minds of men being once roused from their lethargy, and put into a fermentation, turn themselves on all sides, and carry improvements into every art and science. Profound ignorance is totally banished, and men enjoy the privilege of rational creatures, to think as well as to act, to cultivate the pleasures of the mind as well as those of the body.[45]

The passage then ends with a further survey of the benefits of refinement:

The more these refined arts advance, the more sociable men become: nor is it possible, that, when enriched with science, and possessed of a fund of conversation, they should be contented to remain in solitude, or live with their fellow citizens in that distant manner, which is peculiar to ignorant and barbarous nations. They flock into cities; love to receive and communicate knowledge; to show their wit or their breeding; their taste in conversation or living, in clothes or in furniture. Curiosity allures the wise; vanity the foolish; and pleasure both. Particular clubs and societies are everywhere formed: both sexes meet in an easy and sociable manner; and the tempers of men refine apace. So that, besides the improvements which they receive from knowledge and the liberal arts, it is impossible but that they must feel an increase of humanity, from the very habit of conversing together, and contributing to each other's pleasure and entertainment. Thus industry, knowledge, and humanity, are linked together, by an indissoluble chain, and are found, from experience as well as reason, to be peculiar to the more polished, and, what are commonly denominated, the more luxurious ages.[46]

It will be clear that Hume's argument all through the above passage rests on the strength of his generalisations about universal human psychology and about the undifferentiated nature of the social processes he describes. His points of departure are always the undivided categories of 'man' and 'the mind'. In line with this, he represents social progress as a universally beneficial phenomenon, claiming that in commercial society, all the arts advance together under the influence of 'the spirit of the age', satisfying the needs and increasing the happiness of all of those involved in the process. When we examine the passage, however, we find that the ambiguity we have already identified in the first sentence is followed by a further succession of strategic ambiguities, and by conflations and elisions of terminology, which are necessary if he is to sustain his central generalisations without open acknowledgement of the distinctions and divisions that they conceal.

This point becomes clear when, for example, we consider the overtones that are acquired by the term 'refinement' in the course of the passage. As we have already mentioned, at the beginning of the essay, luxury is paraphrased as 'great refinement in the gratification of the senses'; later, 'innocent luxury' is defined as 'a refinement in the arts and conveniences of life'.[47] In between, the connotations of the word 'refinement' are allowed to change and this colours the argument differently in its successive stages. In the discussion of the development of the arts, 'refinement' seems at first to mean simply improvement in the skills necessary in the practice of all the arts. In this sense, in the first paragraph, all the arts are endorsed equally for their effect in stimulating 'industry'. However, 'refinement' is also a term of qualitative distinction between the different arts. This becomes clear in the later paragraphs, in the distinction between 'mechanical' and 'liberal' arts. This distinction is only introduced in the second paragraph, after the general propositions about the 'arts' as a whole have been formulated; and it is re-cast in the third, which begins with a reference to 'these refined arts', and in which the 'refined arts' are presumably – or possibly – set against the 'mechanical arts' mentioned earlier. In the course of the discussion, the initial encomium on the 'arts' of 'industry' has been replaced by a catalogue of the effects of the polite 'arts' of leisure, and yet something of the tone of endorsement of the labour involved in the first has been carried over into the second.

A similar effect occurs with many of the other terms employed in the passage. In psychological and moral terms, 'action' and 'pleasure' are usually treated as being on a par, and as being the opposites of 'indolence'. 'Action' is redefined at one stage as 'business' and 'pleasure' is defined so as to include material rewards, leisure and indeed the 'action' that produces both. In this context, when Hume writes that 'in times when industry and the arts flourish, men are kept in perpetual occupation, and enjoy, as their reward, the occupation itself, as well as those pleasures which are the fruit of their labour', the distinctions between the two initial terms that he employs seem minimal. Both 'action' and 'pleasure' are presented as 'pursuits' to which 'education, custom and example have a mighty influence in turning the mind'. The term 'pursuit' does not necessarily have its modern association with leisure (just as 'business' is not necessarily associated with work). Significantly, however, the economic compulsions that might enforce labour as a species of 'action', even at the expense of 'pleasure', are not mentioned: labour is only mentioned insofar as it purchases 'pleasure' and serves to make a limited 'indolence' 'agreeable'.

At the same time, the focus of Hume's discussion changes at important points in the course of the passage, from consideration of the effects of 'refinement' on individual human happiness to a more traditionally formulated humanist discussion of its effects on 'the nation', a term that conveniently subsumes any potential distinctions between the individuals and professions that make up civil society. This shift of perspective is particularly clear in the paragraph on the manifestations of the 'spirit of the age', where all these potential distinctions are included under the general heading, and Hume writes that 'we cannot reasonably expect, that a piece of woollen cloth will be wrought in perfection in a nation which is ignorant of astronomy, or where ethics are neglected'. As is the case with the other implicit qualifications in his account of society, which we have have highlighted earlier, examination of the formulation that he employs here leads us to recognise that the claims that he makes for the 'indissoluble chain' which links 'industry, knowledge, and humanity' in 'polished' and 'luxurious' ages are extended so as to cover the sense that those things may co-exist in society, but need not necessarily exist with the equality that he suggests for all the men who constitute it.

One of the most striking ways in which Scottish academic studies of society and the economy written after the *Essays* depart from the positions that we have been outlining can be seen in the later writers' investigations of precisely the distinctions and divisions in society that are occluded in Hume's discussions. These investigations characteristically take the form of inquiries into the economic and cultural effects of the division of labour in society, and are crucial to the analyses offered by – amongst others – Adam Ferguson and Adam Smith. Indeed, in some respects the writings of these two men can be seen as dissections of the comprehensive generalisations on which Hume's arguments depend. The most marked examples of this dissection occur in Ferguson's *Essay on the History of Civil Society* (1767). Ferguson does not share Hume's commitment to progress and conviction of the beneficial effects of commerce. Instead he maintains a far more open-ended dialectic between the various forces at work in commercial society. In his argument, as in Smith's, economic development depends upon and promotes the division of labour. At the start of Part IV of the *Essay* he writes, 'the progress of commerce is but a continued subdivision of the mechanical arts', and acknowledges the material benefits that it brings with it: 'By the separation of arts and professions, the sources of wealth are laid open; every species of material is wrought up to the greatest perfection, and every commodity is produced in the greatest abundance'. He then goes on to concede the apparent advantages of the divisions of labour that follow on from these economic divisions, in the 'higher' arts of statecraft: 'The advantages gained by the inferior branches of manufacture by separation of their parts, seem to be equalled by those which arise from a similar device in the higher departments of policy and war'.[48]

His critique of the effects of this process is couched in part in material terms. He suggests that the progress of commerce increases the material differences in the social hierarchy: 'In the result of commercial arts, inequalities of fortune are greatly increased, and the majority of every people are obliged by necessity, or at least strongly incited by ambition and avarice, to employ every talent they possess'.[49] However, more importantly, the *Essay* incorporates an extended discussion of the cultural effects of the divisions of labour that are consequent on the development of commercial society. This is conducted in terms of an assessment of the effects of commerce on the 'capacity' of the nation – a quality judged largely by the degree of martial prowess and civic virtue maintained by its inhabitants. In this discussion Ferguson retains two paradigms of the complete man:

one, the self-sufficient independent savage, and the other the humanist citizen. Both provide critical models, in relation to which all the subjects of the modern state are seen to be specialised, socially constructed products of an economy that they cannot comprehend. In modern society,

Every individual is distinguished by his calling, and has a place to which he is fitted. The savage, who knows no distinction but that of his merit, of his sex, or of his species, and to whom his community is the sovereign object of affection, is astonished to find, that in a scene of this nature, his being a man does not qualify him for any station whatever: he flies to the woods with amazement, distaste, and aversion.[50]

Following on from this, Ferguson inquires into the consequences of the different specialisations on which society depends, as they affect those who participate in them. He presents them as reinforcing in cultural terms the hierarchic economic divisions he has already noted. He writes, for instance, that

Many mechanical arts, indeed, require no capacity; they succeed best under a total suppression of sentiment and reason; and ignorance is the mother of industry as well as of superstition. . . . But if many parts in the practice of every art, and in the detail of every department, require no abilities, or actually tend to contract and to limit the views of the mind, there are others which lead to general reflections, and to enlargement of thought.[51]

He gives some examples, noting that 'even in manufacture, the genius of the master, perhaps, is cultivated, while that of the inferior workman lies waste', and contrasting the 'statesman' and his 'tools', and the 'general officer' and the 'soldier'. Interestingly, he then extends his observation to include his own position as commentator. In doing so he lays out the grounds on which to construct a critique of the 'polite' perspective on society that we have been examining in this essay:

The practitioner of every art and profession may afford matter of general speculation to the man of science; and thinking itself, in this age of separations, may become a peculiar craft. In the bustle of civil pursuits and occupations, men appear in a variety of lights, and suggest matter of inquiry and fancy, by which conversation is enlivened and greatly enlarged. The productions of ingenuity are brought to the market; and men are willing to pay for whatever has a tendency to inform or amuse. By this means the idle, as well as the busy, contribute to forward the progress of arts, and bestow on polished nations that air of superior ingenuity, under which they appear to have gained the ends that were pursued by the savage in his forest, knowledge, order, and wealth.[52]

The passage raises all sorts of points in suggestive terms, without developing them to their potential conclusions. 'The man of science' is presented as being in one sense set apart from the restrictions that affect those engaged in the other 'arts' and 'professions', and as being capable of 'general speculation' in consequence. However, at the same time, his 'profession' is regarded as being as much a specialised product of the division of labour as are those of the men he observes. Despite his apparent detachment from 'the bustle of civil affairs', he is himself involved in producing a commodity for the market, which is consumed by 'the idle' as their contribution to the 'progress of arts'. He is thus fully involved in the deeply ambiguous progress or corruption of society, and his position is open to critique in all directions. The 'science' that he professes is by no means excluded from the debilitating effects of the division of labour, which affect all the branches of social life. The example of Athens is called on as a general confirmation of this, and Ferguson writes, 'the history of this people abundantly shewed, that men ceased to be citizens, even to be good poets and orators, in proportion as they came to be distinguished by the profession of these, and other separate crafts'.[53]

All the elements in Hume's account of the progress of the arts and of the economy are then laid out and opened to critique under the heading of 'Corruption incident to Polished Nations', for the effect they have in separating and individualising the citizens of the state in their pursuit of private ends:

The increasing regard with which men appear, in the progress of commercial arts, to study their profit, or the delicacy with which they refine on their pleasures; even industry itself, or the habit of application to a tedious employment, in which no honours are won, may, perhaps, be considered as indications of a growing attention to interest, or of effeminacy, contracted in the enjoyment of ease and conveniency. Every successive art, in which the individual is taught to improve his fortune, is, in reality, an addition to his private engagements, and a new avocation of his mind from the public.[54]

The effects of this divisive pursuit of self-interest are traced as far as the domain of 'politeness', which Ferguson reads as the triumph of 'effeminacy' and 'interest' over the active and martial virtues, and as the ultimate sign of withdrawal from the public demands of citizenship. He claims that 'the period is come, when no engagement remaining on the part of the public, private interest, and animal pleasure, become the sovereign objects of care'; presents the widespread individual cultivation of 'sensibility and delicacy' as evidence of this development; and insists that 'in this condition, mankind generally flatter their own imbecility under the name of politeness'. Far from being dissociated from the operations of self-interest, 'politeness' is seen as being a particularly pernicious cultural manifestation of that self-interest; far from offering a privileged point of pespective, outside and above specialisation, from which society can be comprehensively observed, it is regarded as being as much the product of specialisation as any other aspect of social life; and far from providing the ultimate vindication of the progress of civil society, its development is presented as being the acme of corruption. Under this scrutiny the divisions and distinctions that are subsumed into the 'polite' perspective on society that is offered in Hume's *Essays* are revealed and laid open to critique.

Notes

1 J. G. A. Pocock, *The Machiavellian Moment: Florentine Republican Thought and the Atlantic Republican Tradition*, Princeton, N.J., 1975, p. 497.

2 E. A. J. Johnson, *Predecessors of Adam Smith: the Growth of British Economic Thought*, New York, 1937, p. 163.

3 E. J. Hundert, 'The achievement motive in Hume's political economy', *Journal of the History of Ideas*, XXXV, 1974, pp. 139–43.

4 *Ibid.*, p. 141.

5 See William Temple, *The Works of Sir William Temple, bart.*, 2 vols., London, 1729. For the influence of Mandeville see T. A. Horne, *The Social Thought of Bernard Mandeville*, London, 1978, esp. chapter 3.

6 References to David Hume's *Essays* throughout are to *Essays: Moral, Political and Literary*, eds. Green and Grose, 2 vols., London, 1898. Here, 'Of Commerce', I, pp. 296–7.

7 John Robertson, 'The Scottish Enlightenment and the limits of the civic tradition', in *Wealth and Virtue: The Shaping of Political Economy in the Scottish Enlightenment*, eds. I. Hont and M. Ignatieff, Cambridge, 1983, pp. 137–78.

8 *Ibid.*, p. 159.

9 Pocock, *Machiavellian Moment*, p. 426.

10 *Ibid.*, p. 493.

11 Robertson, 'Scottish Englightenment', esp. pp. 158–65.

12 Johnson, *Predecessors of Adam Smith*, chapter XIII, pp. 259–80.

13 *Ibid.*, p. 261.

14 See Nicholas Phillipson: 'Towards a definition of the Scottish Enlightenment', in *City and Society in the Eighteenth Century*, eds. P. Fritz and D. Williams, Toronto, 1973, pp. 125–47; 'Culture and society in the eighteenth-century province: the case of Edinburgh and the Scottish Englightenment', in *The University in Society*, ed. L. Stone, 2 vols., Princeton, N.J., 1974, II, pp. 407–48; 'Adam Smith as civic moralist', in Hont and Ignatieff, eds., *Wealth and Virtue*, pp. 179–202.

15 See J. G. A. Pocock, 'Cambridge paradigms and Scotch philosophers: a study of the relation between the civic humanist and the civil jurisprudential interpretation of eighteenth-century social thought', in Hont and Ignatieff, eds., *Wealth and Virtue*, pp. 235–52. Here, p. 240.

16 Phillipson, 'Adam Smith as civic moralist', pp. 179–80.

17 Hume, *Essays*, I, p. 41.

18 *Ibid.*, p. 42.

19 *The Spectator*, ed. D. F. Bond, 4 vols., Oxford, 1965. Here, I, pp. 1–5.

20 The point is discussed in E. A. and L. D. Bloom, *Joseph Addison's Sociable Animal*, Providence, Rhode Island, 1971, pp. 11–28.

21 *Spectator*, IV, p. 306.

22 *Ibid.*, III, p. 19.

23 *Ibid.*, p. 21.

24 *Ibid.*, p. 538.

25 *Ibid.*, p. 539.

26 Hume, *Essays*, I, p. 91.

27 *Ibid.*

28 *Ibid.*, p. 92.

29 *Ibid.*

30 *Ibid.*

31 *Ibid.*

32 *Ibid.*, p. 93.

33 *Ibid.*, pp. 93–4.

34 *Ibid.*, p. 176.

35 *Ibid.*, p. 325.

36 *Ibid.*, pp. 293–4.

37 *Ibid.*

38 *Ibid.*, p. 294.

39 *Ibid.*, p. 295.

40 *Ibid.*, p. 299.

41 *Ibid.*, p. 300.

42 John Barrell, *The Dark Side of the Landscape: The Rural Poor in English Painting, 1730–1840*, Cambridge, 1980, pp. 49–50.

43 Hume, *Essays*, I, p. 300.

44 *Ibid.*, pp. 300–301.

45 *Ibid.*, p. 301.
46 *Ibid.*, pp. 301–2.
47 *Ibid.*, p. 307.
48 Adam Ferguson, *An Essay on the History of Civil Society* (1767), ed. D. Forbes, Edinburgh, 1966, pp. 180–84.
49 *Ibid.*, p. 217.
50 *Ibid.*, p. 181.
51 *Ibid.*, pp. 182–3.
52 *Ibid.*, p. 183–4.
53 *Ibid.*, p. 218.
54 *Ibid.*, p. 255.

10 *John R. R. Christie*

Adam Smith's metaphysics of language

Adam Smith is not a name normally associated with the history of linguistics or rhetoric, nor even with the history of philosophy. He is *par excellence* a case of an author whose intellectual range and diversity, and consequently whose historical significance, has been obscured by his authorship of one great work, *An Inquiry into the Nature and Causes of the Wealth of Nations* (1776), the founding text of classical political economy. Had he never written and published that work, or had he completed his project of a connected history of the elegant arts and sciences, he would probably appear much more regularly than he does as one of the omnicompetent intellects of the European Enlightenment, akin not to Ricardo but to such as Rousseau, Diderot or Hume: less the originary political economist, more the *philosophe*. Only relatively few scholars, Smith specialists or Scottish Englightenment *afficionados*, attempt to comprehend the extent of Smith's work in philosophy, and even so the *telos* exerted by the *Wealth of Nations* remains powerful, for Smith's pre-*Wealth of Nations* work tends to be read for the light it may throw forward to illuminate the text of the *Wealth of Nations* itself. Yet before the *Wealth of Nations* was written, Smith wrote philosophical essays on the external senses and on the histories of astronomy, ancient physics, metaphysics and logic. He wrote also on various kinds of art, on morals and on language, and his educational career demonstrated a long-term commitment to the subject of rhetoric.[1]

Smith's early philosophical and educational interests revolved around topics within epistemology, science and language, and it is his work in these areas which is dealt with in this essay. Most attention is focused on the essay on language, but it is also maintained that a thorough grasp of Smith's views on language requires some notice to be taken of his speculations on epistemology and science. Relevant aspects of his essays on those subjects form a part of the treatment which follows because, as will appear, the essay on language exhibits certain continuities, formal and substantive, with the essays on epistemology and science. These continuities, to mention them merely by way of introduction, include Smith's conception of nature, the unitary model of historical development with which he conceptualises the histories of language and science, and the generic form of conjectural historical analysis utilised by Smith in framing these subjects.

'Conjectural history' is a coinage of Dugald Stewart, used first with reference to Smith's work on language, and to define a kind of inquiry which was, in Stewart's view, 'entirely of modern origin'.[2] It is utilised, wrote Stewart, 'when we are unable to ascertain how men have actually conducted themselves upon particular occasions'.[3] Bereft of particularised historical knowledge, the historian instead will consider 'in what manner they are likely to have proceeded, from the principles of their nature, and the circumstances of their external situation'.[4] Conjectural histories are temporally idealised narratives whose ordering is, strictly speaking, not empirically founded, but entailed by the author's conception of human nature and the probabilities of its response when placed within, and obliged to respond to, an authorially given environment. This environment may be specified in any or all sorts of ways: conceptual, linguistic, political, social, economic. In such narratives empirical historical materials play an illustrative rather than a fundamentally evidential role. Stewart was correct to use the epithet 'modern' to describe this form of historical writing. The high degree of flexibility which the form confers is a measure of that freedom necessary to rewrite and re-evaluate the history of the world as the emergence of the modern. Substantially, its modernity consists precisely in its universalisation and retrojection of the cognitive and affective reflexes of modernity, so that the history it recovers is always of a modernity already prescribed and inscribed within the subject of the historical process, namely man. As Walter Bagehot put it with reference to one example of the genre, and with perfect economy of metropolitan insult, it tells of how, 'from being a savage, man rose to be a Scotchman'.[5]

Adam Smith used conjectural historical analysis throughout his intellectual career. One of his earlier works, 'The principles which lead and direct philosophical enquiries: illustrated by the history of astronomy',[6] is worth preliminary attention because it eventually focuses on a central difficulty of language, and does so in a way which indicates Smith's own rhetorical sensitivity to the field of language in general. The 'History of astronomy' has an epistemological aim. It is designed to show the conventional nature of scientific explanation, insisting that all systematic theories, such as the solid spheres of the ancients which move celestial bodies, or the vortices of Descartes which perform the same function, were products of human imagination, not approaching anything like a true or realistic representation of nature. These explanatory mechanisms conformed to the criteria of the imagination (unity, simplicity, harmony), rather than to reason and nature, and were compelled from their authors by psychological necessity, by man's passional nature, which will not let him rest content without a coherent and systematised representation of the natural world. Smith's radical de-emphasis of terms of reason, truth and reality in relation to natural science, largely unparalleled until our own time, is highlighted as the essay draws to a close.

> and even we, while we have been endeavouring to represent all philosophical systems as mere inventions of the imagination, to connect together the otherwise disjointed appearances of nature, have insensibly been drawn in, to make use of a language expressing the connecting principles of this one [the Newtonian system], as if they were the real chains which Nature makes use of to bind her several operations.[7]

at a superficial level, this sentence is straightforward enough. Smith did not wish his laudatory account of the Newtonian system, which stressed its profound explanatory coherence, to subvert the leading lesson of the essay. But the sentence demands more understanding – 'we . . . have been insensibly drawn in to make use of a language' which expresses reality. Obviously enough, Smith has *not* 'been insensibly drawn in' to linguistic realism, but has artfully adopted it, the better to enforce his lesson. This lesson is not simply that scientific discourse is radically conventional in nature, but that even despite this philosophical awareness, our utterance, in its slippage towards realistic expression, tends to forget that awareness. Philosophical awareness is not enough in itself, but requires this rhetorical shock to warn of a linguistic threat which undermines philosophy's teaching concerning error. Set within this kind of reading, the precise form of error also becomes at bottom linguistic. It is referential in nature. The systems of science are not to be taken for literal descriptions which refer to actual powers, forces and mechanisms, really existing in nature. They are instead associative chains of ideas, projected from within human imagination onto the external world. As such their referential status with respect to the external world is entirely dubious. The discourse of science, pre-eminently the most literal and empirically referential of all discourse, turns out to conform, in definitional terms, to fiction. It is a field of indeterminate reference. And it would seem that, precisely *because* this problematic nexus is perpetually linguistic in nature, its writing out requires a strong *rhetorical* strategy. Smith's deconstruction of science, which approximates science to fiction within a universe of imagination, cannot persist without a rhetorical embodiment. Rhetoric is the necessary underwriting of philosophy if philosophy is to install itself as consistent awareness against the desire of language for the real.

The mode of 'rhetoric' appearing here in Smith's text is as a means of orientating the reader, and it operates through a concealed mimesis followed by its revelation. Briefly, Smith, without any signalling, changes the register of his writing so that it mimics the error which is soon to be brought to the reader's attention. The error is pinned by revelation of the rhetorical stratagem which mimics it, and within this relatively short compass the reader has been radically reorientated, or more strongly, recreated. Henceforward the philosophical reader of science operates with an additional and crucial level of linguistic consciousness, is a new kind of reader. In general terms, a similar thing happens by the end of Smith's conjectural history of language, if not by the same means: that is, the historical analysis of language itself promotes a novel consciousness of its subject as unsettling as that which Smith performs with regard to science. Enough has now been said of that treatment of science to make it clear that its author, who wrote what we would now call a 'philosophy of science', wrote nonetheless with an acute consciousness of both the difficulties and resources of language, and did so in a way which, befitting somebody who taught rhetoric as part of philosophy, did not marginalise, or trouble to disguise or suppress his rhetorical skills and usages.

Smith's essay, 'Considerations concerning the first formation of languages, and the different genius of original and compounded languages', first took shape, not in its fully worked and published form of 1761, but as a lecture in his series of thirty lectures on rhetoric and belles lettres.[8] Smith had lectured on these subjects from 1748, to audiences in Edinburgh, then to his advanced moral philosophy class at the Unversity of Glasgow. The essay of 1761 is, it is presumed, an elaborated and considerably expanded version of the lecture, some aspects of which may well have featured from the inception of Smith's rhetoric course in 1748. In his correspondence, he names a work of 1747 by the Abbé Gabriel Girard, *Les Vrais Principes de la Langue Francoise* as that 'which first set me a-thinking on these subjects'.[9] As the most recent editor of the essay on language notes, it may well have been the Abbé's propensity to see the generation of grammar in terms of mental actions which appealed to Smith.[10] But if some version was delivered as early as 1748, then certain features were probably added in the 1750s. Smith also instanced the grammatical articles of the *Encyclopédie*, though with the perhaps dismissive term of 'entertainment' attached to them.[11] More suggestively, however, the essay takes up crucially the problem of language left hanging fire by Jean-Jacques Rousseau in his *Dissertation on the Origin and Foundation of the Inequality of Mankind* of 1755.[12] Smith's developing speculations on language were likely to have been moulded around his progressive encounters with these French writings, particularly those of Rousseau. Given this sequence, the question of assigning a definitive dating to the composition of the essay becomes fruitless if not meaningless. In particular, its temporal relation to the essay on the history of astronomy, also composed in the 1750s and with which it exhibits certain parallels, remains finally undecidable.

The essay on language was itself a sign of the growing philosophical attention to language evident throughout much of the European Enlightenment from 1750 onwards. This is an episode largely neglected by conventional history of philosophy, and conveniently forgotten by contemporary linguistic philosophy. Only recently, through the work of Hans Aarsleff, has it started to receive the attention it merits.[13] At first sight, Smith's essay would seem to differ little in its preoccupations from those of other philosophical writers of the time, but a closer examination shows that Smith's perception of language and its problems was less than congruent with those of his contemporaries. Three related problems informed much work on language between 1750 and 1780. These were the specific origin of language, in terms of its derivation from prelinguistic sounds or gestures, or indeed from God; the priority relations between language and society, and the natural or artificial status of language. Elements and assemblages of these problems motivate much of the work of Condillac and Rousseau in France, Herder in Germany, and Lord Monboddo, writing a decade or so after Smith in Scotland.[14] It may be that Smith's views on these matters are technically deducible, but they did not appear in the essay as *problems*, stated and solved as such. For Smith, such cruxes appear to be either minor, or possibly false. At any rate, it is clear from the essay's subject-matter that his phrase 'first formation of languages' does not equate with the originary question as stated above, that the priority of language to society or *vice versa* was less than germane, and that he viewed language as artificial in so far as it was obviously and ascertainably the product of human 'art', but that equally this 'art' was natural to man. The avoidance of these cruxes stems from the distinctive framework of Smith's approach. The keyword in his title is 'formation'. He was not concerned with a specific and originary derivation for vocal and written signs, but with the way in which such signs subsisted as categorical members of a structure or system: language as attaining that form we call grammar. Put briefly, the historical lineage of Smith's essay descends from the universal grammars of the seventeenth century, through the epistemological turn given to grammar by Girard and Condillac, and finally to Rousseau's perception of what was problematic in this more recent work, namely the difficulty of envisaging how the primitive mind could have formulated certain rudiments of speech, given their highly abstract or metaphysical nature.[15] Smith treated this legacy by historicising it, which was to abandon the rationalistic universalised grammar typified by Port-Royal, and the elaborate epistemological devising of Condillac. Instead, he submitted grammar to a series of historical hypotheses, and it is this which tends to dismiss any preoccupation with origins from the essay. As Smith will tell us, grammar is always present beyond the primitive stage of first denominations of things and actions. But because it is subject to historical development, its forms are always originating. History renders grammar plural and various. Its uniformity only exists in its being the condition of meaningful signification. Put another way, grammar for Smith is what enables language to *work* (a term to be used advisedly when interpreting the theorist of labour), but the means of that workability, the systems men use to be able to signify, exhibit developmental change through history.

Smith initiated his speculations on language by imagining two savages, living in a lightly-sketched background of caves, trees and fountains.[16] By uttering sounds, they attempt to communicate their wants to one another and so denominate their subsistence: the cave which shelters, the tree whose fruit feeds, the fountain which slakes their thirst. This is as close to an origin for language that Smith comes, and can be described as being both social and material, in that Smith represents it as occurring in the minimum community of two persons, bound by material necessity. With no hint that this conception might be problematic, Smith moves rapidly forward. As experience of the environment is enlarged and other caves, trees and fountains are encountered, the savages inevitably recollect their initial encounters with examples of these objects, and use for these new samples the initial denominations of the original encounter.[17] With this activity the grammatical formation of language has begun. The first denominations by which the savages named the objects constituting their world were all the proper names of individuals. As experienced enlarged, perception of similarity and the faculty of memory transformed the proper to the common, 'Mankind', wrote Smith, 'are naturally disposed to give one object the name of any other, which nearly resembles it, and thus denominates a multitude, by what was originally intended to express an individual'.[18]

This activity was the basis for the primary arrangement of objects into classes and types through which mankind began to grasp the external world, an activity, Smith continued, 'which the ingenious and eloquent M. Rousseau of Geneva finds himself so much at a loss to account for the origin'.[19] With examples from the speech of children and adults which underpinned his contention of the naturalness and prevalence of the linguistic habit which generates generalised substantives, common nouns, Smith felt he had eliminated Rousseau's source of bafflement.[20] However, Rousseau's problem was solved only by altering its premises. Rousseau had emphasised the generalised, abstract and therefore difficult nature of terms of genera and species.[21] He had then set up an insoluble circularity by claiming that generalisation itself depends upon prior possession of the appropriate words, but that the words themselves only become possible by the power to generalise.[22] Each is the prior enablement of the other. Smith only squared this circle by cutting out the power to generalise, demonstrating it as a common natural ability, and giving it the status of prior cause of general terms. Even so, it is clear that

Smith took the intent and the language, if not the form, of Rousseau's problem seriously. Rousseau had perceived the difficulty of conceiving the formation of language when even basic elements of a linguistic system indicate high degrees of abstraction and generalisation. These terms, together with 'metaphysics', appeared in Rousseau's writing,[23] and Smith's essay, which used and developed the same vocabulary, can be profitably read as a historicised meditation on, and vindication of, Rousseau's briefly and forcefully expressed propositions.

From the invention of proper and common substantives, Smith turned to adjectives, that class of words which, by expressing the quality of an object, serves to distinguish particular objects from others in the same class. Simple in function, adjectives are implicated in an additional level of metaphysical complexity when compared with generalisation to common names. An adjectival invention requires prior comparison of many objects, observing their similarities and differences. It presupposes a class of things to which it is applicable, and is therefore a very general term. At the same time, its invention has required prior abstraction because it implies the separability of quality from object. 'The invention, therefore, of even the simplest nouns adjective, must have required more metaphysics than we are apt to be aware of', as Smith put it.[24] Adjectives, terms which serve to individuate or differentiate, incorporate generalisation of similars and abstraction of quality from substance. Prepositions take differentiation a stage further, functioning to distinguish objects when distinction by quality, adjectivally, is inadequate or inappropriate. They perform this function by denoting a relation subsisting between objects, specifying it concretely and correlatively. This involves even more generalisation and abstraction than the case of adjectives. A relation is 'a more metaphysical object than a quality. Nobody can be at a loss to explain what is meant by a quality; but few people will find themselves able to express, very distinctly, what is understood by a relation.'[25] Unlike qualities, relations are never the objects of our external senses. Before the invention of a word instituting the sign of a relation, and nothing but the relation, the relation must have been conceived in abstraction from the related objects, because nothing whatever in those objects, or their ideas, could begin to signify the relation. Again, prepositions, as very general terms, must have required prior distinctions and comparisons, often subtle and complex.

> [T]he preposition *of* [denotes a relation] of a very metaphysical nature. The preposition *of*, denotes relation in general, considered in concrete with the co-relative object. It marks that the noun substantive which goes before it, is somehow or other related to that which comes after it, but without in any respect ascertaining . . . what is the peculiar nature of that relation. We often apply it, therefore, to express the most opposite relations; because, the most opposite relations agree so far that each of them comprehends in it the general idea or nature of a relation . . . and so far as any particular relation is collected from such expressions, it is inferred by the mind, not from the preposition itself, but from the nature and arrangement of the substantives, between which the preposition is placed.
>
> What I have said concerning the preposition *of*, may in some measure be applied to the prepositions *to, for, with, by*. . . .[26]

The metaphysical incorporation marked by prepositions is then even more considerable than adjectives. They indicate a level of severe mental complexity underlying their formulation. Ask a man to describe the relation expressed by *of*, and 'if he has not beforehand employed his thought a good deal upon these subjects, you may safely allow him a week to consider of his answer'.[27]

The last two parts of speech discussed by Smith are verbs and pronouns. Verbs are necessarily coeval with the beginnings of formed language, for all affirmations require them. The earliest and simplest verbs are impersonal. They 'express in one word a complete event, preserv[ing] in the expression that perfect simplicity and unity which there always is in the object and idea; and suppose no abstraction or metaphysical division of the event into its several constituent members of subject and attribute'.[28] Verbs increase their metaphysical element in shifting from impersonal to personal. Smith imagined a group of savages initially using an impersonal verb of the form *venit* (it is coming), as in *pluit* (it is raining), to denote only one particular coming, not coming in general: the coming, in this case, of a lion. 'This word thus expressed the complete event, without the assistance of any other.'[29] As time wore on, they would have generalised the range of *venit* to include the coming of any terrible object. The signification has thereby shifted from one particular coming to the coming of one particular kind. However, the increased utility of more generalised application involves loss of reference in terms of naming what precisely is coming, thus to signify a particular terrible coming a substantive would have to be added to the verb. The savages must shriek *venit leo* where before they shrieked *venit*, to distinguish it from *venit ursus, venit lupus,* etc. In shifting thus from impersonal to personal, the verb has done so in terms of increased metaphysics. The metaphysical component here is that of analysis. The impersonal verb expressed the event synthetically, that is completely and instantaneously; the personal verb expresses it analytically, dividing subject and action.

According to Smith, verbs commenced their historical career as synthetic impersonals, and continued it as analytic third person affirmations, but, simply because verbs had the potential for affirming of two other persons, ways of expressing those relations became necessary. This was his introduction to pronouns, the furthest exemplars of language's metaphysicality. Of *I* and *you* he reflected, 'though custom has now rendered them familiar to us, they, both of them, express ideas extremely metaphysical and complex'.[30] The discussion of *I,* the pronominal sign of the self, is worth special attention. Whatever speaks calls itself *I,* so it is a very general word, 'capable of being predicated, as the logicians say, of an infinite variety of objects'.[31] Yet it differs utterly from all other general words because 'the objects of which it can be predicated do not form any particular species of objects distinguished from all others'.[32] The immediately obvious objection that *I* is predicated of a particular species, the human, misses the point. '*I* does not, like the word *man,* denote a particular class of objects separated from all others by peculiar qualities of their own. It is far from being the name of a species, but, on the contrary . . . it always denotes a precise individual, the particular person who speaks'.[33] For Smith, the logic of *I* appears as an extremely complex signification, uniting within one sign the opposites of utter individuality and generalisation. Seeking to express it in other terms, one finds oneself forced towards oxymoron. In Smith's analysis *I* is a universal singularity.

As such it may be taken as an emblem of Smith's metaphysics of language. The historical line he traces, from proper to common nouns, from adjectives to prepositions, from impersonal verbs to the first person pronoun, is a steadily mounting trajectory of metaphysical complexity, as the sequential invention of basic parts of speech requires progressively more difficult acts of generalisation, abstraction and analysis. Language becomes metaphysical precisely in proportion to its attainment of grammatical form. Grammar simply is metaphysics. This is ultimately what delivers Smith's rule of linguistic historical development. It is commendably straightforward. The more metaphysical freight a grammatical category, or any individual word within a category, has to bear, the more difficult it was to invent, therefore the later its invention. (See Figure 10.1 and 10.2 which represent this graphically for nouns, adjectives and prepositions, and for verbs and pronouns.)

In Smith's delineation, language performs the simple task of referring to objects, events and their qualities and relations. Its means of so doing are, as we have seen, decidedly complex. Moreover, these means are subject to considerable variation through time. The character, or in Smith's usage the genius, of language is subject to historical development, and this topic formed a substantial part of his essay. So far this treatment has taken the majority of its examples from compounded, that is modern, language. But as early as the

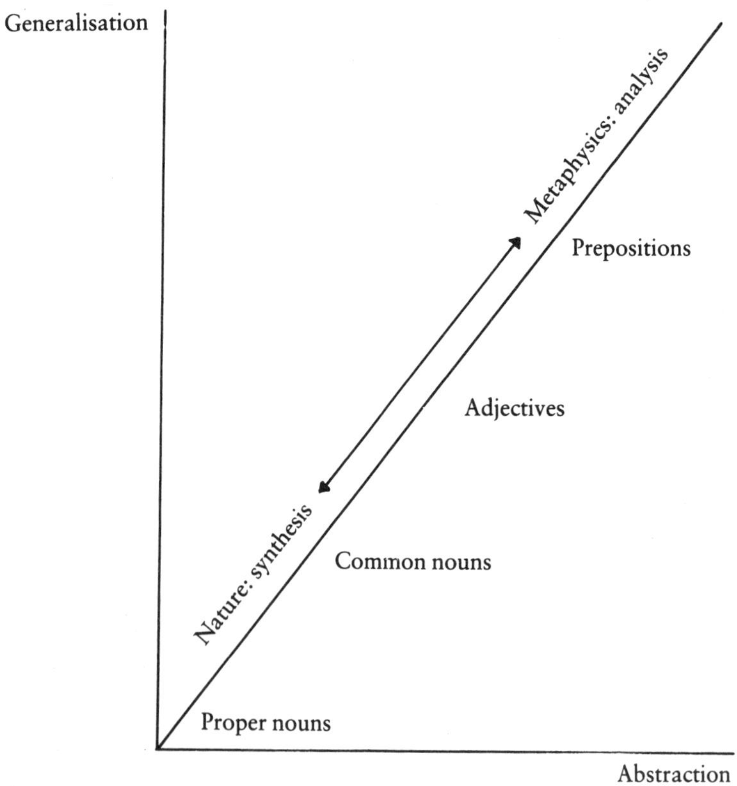

10.1 Nouns, adjectives, prepositions

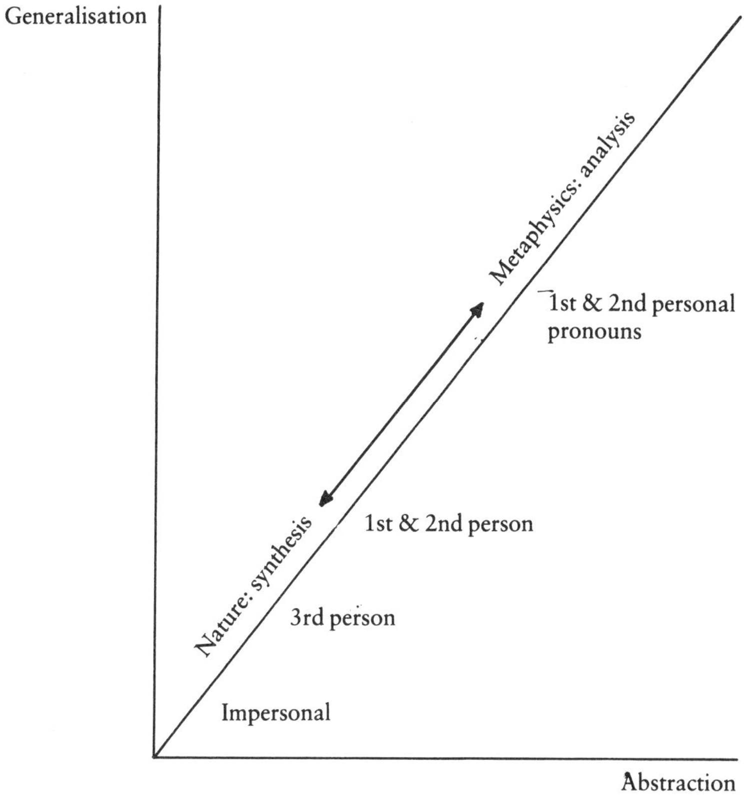

10.2 Verbs, pronouns

adjectival stage of development, Smith introduced considerations of classical language – primitive language in his designation. Having discussed the generality and abstraction inherent in adjectival invention, Smith introduced another form of denoting qualitative differention. This was by varying the substantive itself, the main example being Latin gender terminations. This mode of qualifying lacks the degree of metaphysicality possessed by modern qualification.

Both sex, and the want of all sex, being naturally considered as qualities modifying and inseparable from the particular substances to which they belong, it was natural to express them rather by a modification in the substantitve, than by any general and abstract word expressive of this particular species of quality. The expression bears . . . a much more exact analogy to the idea or object which it denotes, than in the other [the modern expression]. The quality appears, in nature, as a modification of the substance, and as it is thus expressed, in language, by a modification of the noun substantive, which denotes that substance, the quality and the subject are, in this case, blended together, if I may say so, in the expression, in the same manner as they appear to be in the object and the idea.[34]

Genders were able to cope with the two most important qualifications of substances, animate or inanimate, male or female, without any adjectival addition.

A congruent construction was placed by Smith upon prepositional invention. Primitive language used cases to express relations later given prepositional form. *Fructus arboris* does not express the relation of objects by a word denoting relation and nothing other. The relation is expressed with no prior effort of abstraction, but 'as it appears in nature, not as something separated and detached, but as thoroughly mixed and blended with the co-relative object'.[35] The equivalent analysis for verbs and pronouns has already been foreshadowed. Impersonal verbs such as *pluit* express the event in a fashion closer to nature and our perception of it than later, personalised affirmations such as *tempestas est pluvia*. The development of persons in verbs, expressing person by termination, precedes those most metaphysical words, the first and second personal pronouns. The pattern then is clear. Gender (and number) forestalls adjectives, declension forestalls prepositions, conjugation forestalls pronouns. Since these grammatical categories represent stages on the ascending metaphysical scale then the inflected grammatical systems of primitive language constitute a historical deferral or evasion of much metaphysics.

It has so far been clear that language develops in the direction of metaphysics. The terms of Smith's analysis of primitive language have now also made it clear that if at any time language is headed towards metaphysics, it is also receding from nature. This needs careful specification, given that this analysis has broached the topics of the origin and naturalness of language only to maintain Smith at some distance from them. Smith did not cleave to any notions of an original, perfect and natural language from which all language is a subsequent degradation. Still less did any hint appear of an original divinely-inspired Adamic language, where signifier and signified coincide, where sign perfectly expresses and thereby commands essence, such as can be found in revoltionary England over a century earlier.[36] Nonetheless his analysis of classical language undoubtedly presents a preference for the early grammatically inflected systems, and this preference is to do with how their forms of expression stand with regard to both nature and our perception of it. The grammatical strategies of inflection were closer to nature for Smith in that they possessed a particular one-to-one correspondence with their referents, the objects and ideas they signified. The form of the word mimics that for which it is a sign, so that part at least of its signifying work is performed by its being a formal analogue of its referent. There is a general historical point to be made here, concerning John Locke's basic doctrines of language, and any presumption that these doctrines were thoroughly paradigmatic for eighteenth-century linguistic theory. Locke's leading notions, of the arbitrary relation of signifier and signified, and of words as signs for ideas, not objects, had little regulative force for Smith. Smith, as material already quoted indicates, wrote of words as standing equally for objects and their ideas, and saw inflected language as characterised by a certain natural correspondence to the world. There is an ultimate condition for this correspondence, but it moves us away from Locke in the direction of Berkeley, and will be better dealt with in an altered interpretive context which is to follow.

Before moving to that context, there remain significant elements of Smith's history of language to be considered. The crucial historical passages of language are from the early savage to primitive, and from primitive to compounded. A definitive element in these movements, the drift from nature to metaphysics, is already present, but of itself does not fully describe the crucial transitions, nor fully characterise the three stages. The very earliest language Smith envisaged as composed of proper names and impersonal verbs. Impersonal verbs might even here be called the proper names of events and actions. Such language, however, rapidly becomes unworkable. If every single item has its own proper name, then the number of words required rapidly accelerates beyond human capacity to retain and employ. We are in linguistic territory adjacent to that occupied by Funes the Memorious.[37] Men are therefore obliged to institute words which denote not events, but the elements composing them. Expression of any event thereby becomes more complex, but the system of expression is conversely simplified, for the elements of language are now repeatable, transferrable from one situation to another. Much fewer in number, they can be memorised and used.

These elements subsist as members of the sophisticated additional structures of inflected grammar. They incorporate some metaphysics, but the grammar is simply retainable, by rhyme, and the metaphysics is not so considerable as to disallow words from performing as analogues of nature and perception.

As language undergoes the transition from primitive to compounded, inflected to non-inflected, classical to modern, its metaphysical incorporation increases dramatically, with preposition and pronoun progressively in the room of inflections. According to Smith, this process occurs through the mixing of linguistically different nations, and the consequent difficulties of adapting different inflections to one another. Existing in a state of linguistic perplexity, one group would tend to resist alien declension by replacing cases with prepositions. Thus modern Italian has *di Roma,* deriving from *de Roma*, the Lombard's attempt at prepositional substitution for the dative of *Roma*. 'The words,' Smith wrote, 'are, in a great measure the same as before; but the grammar is entirely lost'.[38] Analogous replacements occurred with verbs, where, beset with unintelligible conjugations, the Lombard recruited the device of auxiliary verbs, together with greater use of pronouns. With auxiliaries supplying the active and passive voices, and pronouns the persons, the grammar of conjugation became largely supplanted.

We have now arrived at modern, compounded language. Smith seemed to characterise it in terms of 'lost grammar', but this was not meant in any absolute sense. Grammatical inflection has been lost, but grammatical structure has in some way to be retained, or else the work of signification and reference will run awry. What has happened is that the grammatical work performed by inflection is now carried on by other means. This means Smith designated as compositional; as inflection decreased, grammatical relations had to be increasingly indicated by composition: a greater number of constituents in a more precise ordering of words within sentences. Historically speaking, modern language is the result of solutions to the problems of linguistic mixing. Inflection was the casualty of these encounters. The high degree of metaphysicality exhibited by modern language is the specifically linguistic entailment of the loss of inflection. This analysis led Smith to the formulation of his central historical maxim concerning language. It serves to characterise any language advanced beyond the early savage stage of discrete and unrelated proper names.

> In general it may be laid down for a maxim, that the more simple any language is in its composition, the more complex it must be in its declensions and conjugations; and on the contrary, the more simple it is in its declensions and conjugations, the more complex it must be in its composition.[39]

Any language beyond the early savage must embody a *quantum* of complexity in its system. That complexity appears in two ways only: along the axis of inflection and along the axis of composition. Historically, the graph of development declines from the inflection axis, inclines to the compositional. (See Figure 10.3). The higher the language is on the graph, the later and more metaphysical is its character, for metaphysics, equally as composition, follows from the loss of inflection. Latin, compounded of Greek and Tuscan, has less inflection than Greek, Italian and French less than Latin. English, a

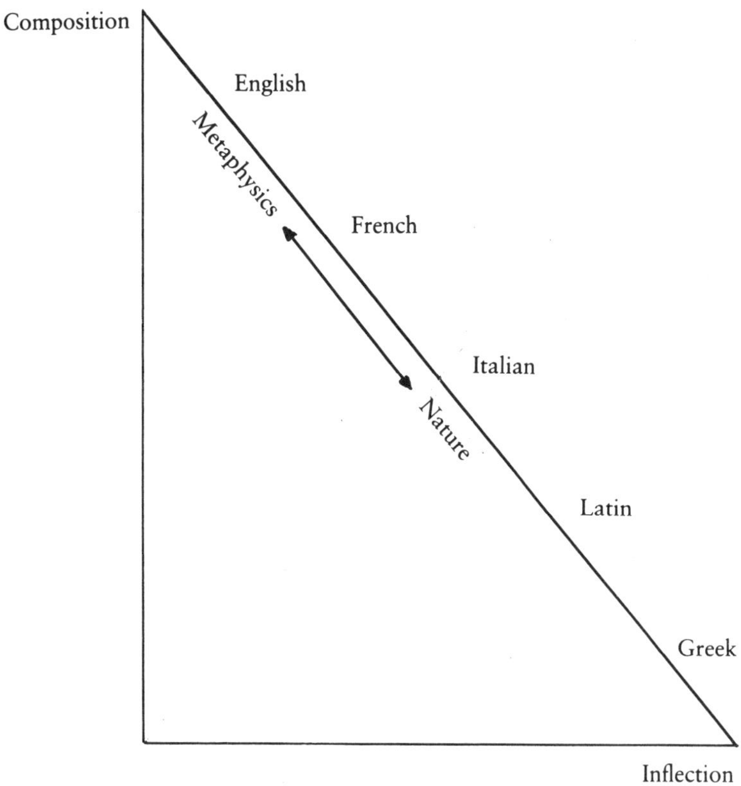

10.3 Language

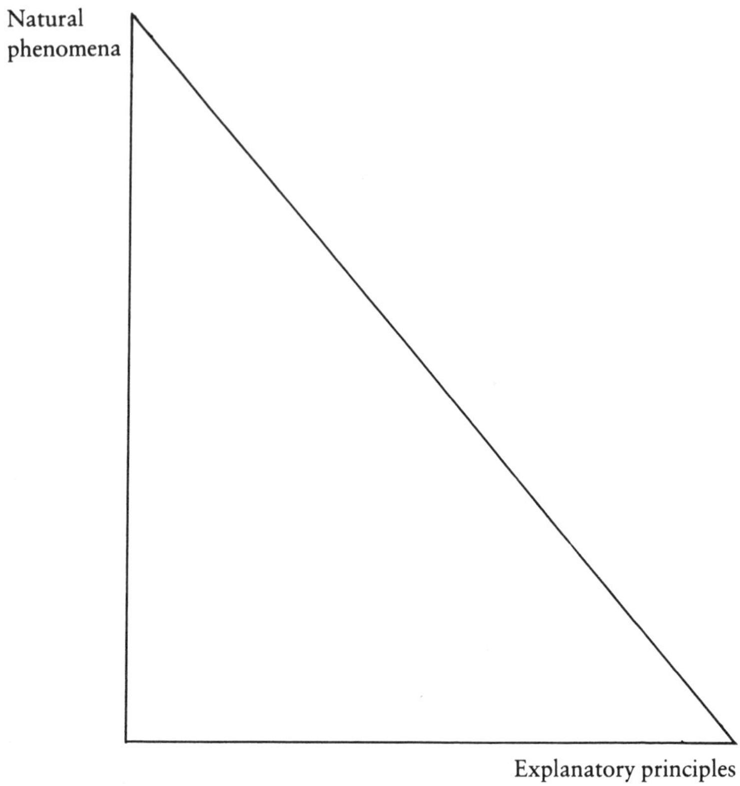

10.4 Scientific systems

compound of Saxon and Norman French, is the most compositionally complex of all languages.

At this point Smith introduced the controlling model of his analysis, though additional evidence is required to maintain the model's controlling status. This evidence comes from the 'History of astronomy', where Smith specified the identical model as a condensed heuristic for his descriptions of how the explanatory systems of natural science develop (see Figure 10.4), and it appeared in attenuated form even in the notes taken by the Glasgow students of the language lecture in his rhetoric course.[40] Each of the crucial transitions of language was undertaken with the motive of simplification. Savages simplified their infinitely expanding denominations with inflection. Lombards simplified the difficulty of coping with alien inflections by using prepositions and pronouns. The metaphysical and compositional complexity which ensued were not the intentional result of willed actions. This is the linguistic version of the historical law of heterogeneity of ends (unintended consequences), much used throughout the whole genre of conjectural history. Smith derived his representation of simplifying development from the history of technology. As language is simplified 'in its rudiments and principles, just as in proportion it grows compositionally more complex, . . . the same thing has happened in it, which commonly happens with regard to mechanical engines'.[41] When machines originate, their principles are complex. Each effect has its own particular motional principle. Machines are developed and improved with the realisation that one principle of motion can

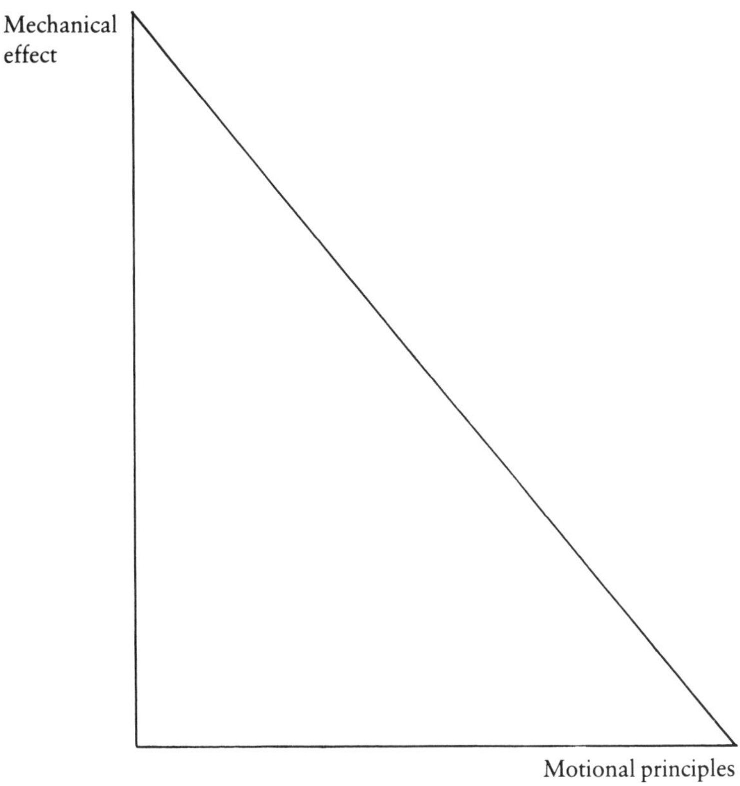

10.5 Machines

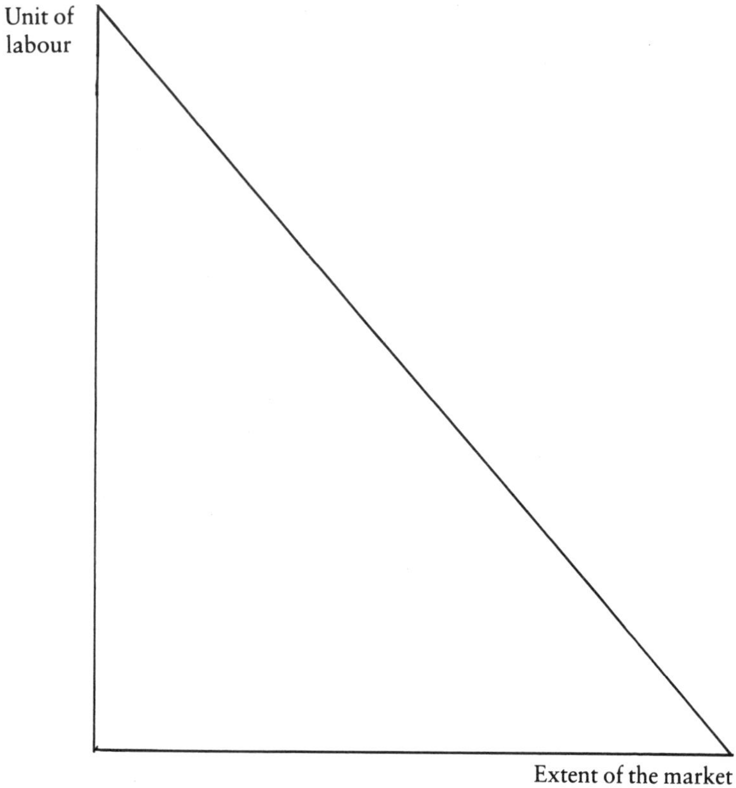

10.6 Labour

supply several types of motion. So are derived machines of mechanically complex effect which are yet essentially simple in the design of their principles (see Figure 10.5.) This model, of simplicity at one end of a system correlated with complexity at the other, is a compression of Smith's extended analyses of both science and language (Cf. Figure 3, 4 and 5).

It may also be said, finally giving way to the temptation to glance forward to the *Wealth of Nations,* that the model provided Smith with the structural and dynamic pattern for his most famous theory. This is the correlation of the division of labour with the extent of the market at the opening of the *Wealth of Nations,* where Smith advances the view that the division of labour increases as the market extends. The unit of labour, in other words, becomes smaller and more simple in proportion to the market's becoming larger and more completely diversified. (See Figure 10.6.) From this latter perspective, Smith's history of language becomes a history of the division of referential labour and its effects upon the nature of language. Inflected words perform a multiple and various referential work which gives way historically to the more specific and singular work of reference performed by modern words in their compositionally extended grammatical framework. Conversely and perhaps more interestingly, this formulation could be inverted to produce the claim that Smith conceived of economic development and its socio-political effects through a model developed in his study of the systematic signification inherent in science and language.[42] Whether one represents Smith's linguistics as a kind of historical materialism or his historical materialism as a kind of semiology matters less, however, than what has now emerged. Smith's treatment of three major elements of human existence, the cognitive, the linguistic and the material go forward in the iconic form of the machine. The machine attains the status of icon in his work, not in its being a pervasive metaphor, but because it is a central sign which itself constitutes the system by which are constructed the signs of the subjects with which he dealt. The historically extended application of this icon was a significantly novel form of cultural perception. Smith was not merely saying that science and language were machines for understanding and expressing the world, but that the development of science and language can be modelled as the development of machines. The diachronic feature means that the machine icon is not just usable for a static snap-shot, a character-study. Generalised, it produces a potentially total thesis concerning the developmental form of all our species-activity. It is thus the mobile image of human history.

The values Smith attached to the forms produced by history emerge strongly in the essay on language. Machines get better in their simplifying development, and so too do scientific explanations, as the 'History of astronomy' makes clear. But the situation is quite the opposite for language. Smith was perfectly explicit in his preference for classical, inflected language over modern, and he portrayed this preference in straightforwardly aesthetic terms. Modern language generates more words, so that one often has to use several for classical language's one. Modern languages are less euphonically agreeable, for the terminations given by declension and conjugation lends a 'sweetness and variety' unavailable in the modern.[43] Modern language suffers greatly from the constraint imposed by strict ordering of words within sentences, whereas the classical has much greater freedom to place words in their most aesthetically euphonic ordering without disturbing their mutual reference, which is taken care of by inflection. Smith made this point by quoting one of Milton's literal translations of Horace.

> Who now enjoys thee credulous all gold,
> Who always vacant, always amiable
> Hopes thee; of flattering gales
> Unmindful —

'Impossible to interpret by any rules of our language,' he commented.[44] Smith's aesthetic valorisation of the classical contrasts abruptly with his deprecation of the modern, which is now a domain characterised by 'prolixness, constraint and monotony'.[45] 'Lost grammar' is also lost beauty.

The linguistically modern in Smith seems then the scene of loss, given that its entailment, a more metaphysically embedded language, is never seen as any sort of compensation. However, Smith's closing his essay with this strongly aesthetic category of valorisation must initiate some questioning in juxtaposition with the issue of metaphysical modernity. This is because metaphysics, in the course of the essay, is present as an ever-increasing form of *cognitive* estrangement. It removes language from that conformity to nature and human nature which it had previously possessed. The 'natural' expression of impersonal verbs was earlier contrasted by Smith with the 'artificial' and 'imperfect' expression introduced by personalised verbs, even within the domain of the classical.[46] What has happened, therefore, is that although Smith has avoided the polarity of 'natural' and 'artificial' as the subject of the essence of language, the polarity has been reinscribed in historical terms as a developmental subject of linguistic history. It appears first in the essay in relation to Smith's epistemological preference; he obviously judges synthetic expression as possessing more fidelity to nature than later analytic expression.

Why then is the essay's closure stated in aesthetic terms? One simple answer might be that for Smith the epistemological and the aesthetic are not properly separable. Synthetic forms of language are related to the objects they express in terms of fitness, an appropriateness of means to ends, and this is an aesthetic relationship. Lack of fitness does not however appear in Smith's complaint against the modern, where the catalogue of absence lists economy of expression, freedom of composition and phonic harmony. Some other answer has to be sought. As it stands, the aesthetic consideration which is the essay's conclusion and the final closure against the modern, radically occludes the essay's main epistemological drive. This, then, is its function, which it performs with great rhetorical effect.

To continue this critique, we need to know the necessity for Smith's device, particularly so when other signs of a troubled text occur precisely in relation to this question of the correspondence of inflected synthetic expression to nature:

> The quality appears, in nature, as a modification of the substance, and as it is expressed, in language, by a modification of the noun substantive, which denotes the substance, the quality and the subject are, in this case, blended together, if I may say so, in the expression, in the same manner as they appear to be in the object and the idea.[47]

For Smith, this was an abnormally hesitant and clumsy way of saying that gendered words formally correspond to their referents. The laborious construction and numerously punctuated pauses of the sentence make it begin to doubt its own assertion, a doubt which is then substantially signalled by the address to the reader, 'if I may say so'.

Why he may say so, and the simultaneous inadmissibility of so doing, are things not readily apparent in the essay on language, although certain of its aspects may point in the appropriate direction. It is now clear, for example, that despite a surface clarity about the relation of language to nature (its early grasp, its later loss), Smith's text could not eradicate, and finally sought to forget, an uneasiness concerning that relationship. The trace to follow is the ever-problematic keyword, *nature,* and it is therefore fortunate that in another early epistemological text, 'Of the external senses', Smith left us his only extended conception of what nature might be.[48] 'Of the external senses' is reliant on Berkeley, particularly the section on the visual sense, which draws heavily upon Berkeley's *New Theory of Vision*. Smith was evidently taken with Berkeley's notion of a language of nature, and wrote of nature's language, the language of vision, as having 'a fitness of representation, an aptitude for signifying the precise things it denotes, much superior to that of any artificial languages which human art and ingenuity have ever been able to invent'.[49] Although Smith used Berkeley's discussion of human and natural language to further his discussion of the relation between visible and tangible objects, natural and artificial language are nonetheless contrasted. In the natural languages of sight and touch, Smith discerned 'some affinity or correspondence' between visual and tangible signs which was not present between the signs of human language and their referents.[50] Nonetheless, Smith proved unable to let go of Berkeley's notion, and it reappears, pushed byond Berkeley, less as a contrast than as a comparison. 'In this language of Nature, it may be said, the analogies are more perfect; the etymologies, the declensions, and conjugations, if one may say so, are more regular than those of any human language'.[51] A situation is hereby reached where nature is not only language, but inflected language. It is not then merely the case that Smith following Berkeley, could conceive nature as language, but more exactly it could be thought of as *grammar*. On this presumption therefore, the condition of inflected language's ability to express nature in terms of correspondence is simply the relation of identity between them. Grammar is already inscribed within nature and inflected, synthetic expression can only be nature's faithful mirror. It should be suggested therefore that this unstated identity is the most plausible root of Smith's uneasiness with this theme in the history of language. In that essay, the collapse into one of the separate categories of nature and grammar is inadmissible because it is precisely the widening gulf between them, named metaphysics and coeval with grammar, whose history was being written.

The responsiveness of Smith's writing to this kind of Derridan reading encourages further questioning. The Derridan mode has revealed Smith as a rhetorician, or perhaps a meta-rhetorician, who had a deeply grammatological perception, but who was unable to integrate this perception structurally within his philosophical historiography, where it remained as a latent troubling of the text. The other mode of reading which immediately recommends itself along this critical path is that formulated by Paul de Man in the opening chapter of *Allegories of Reading*, for de Man raised here exactly this question of the relationship between grammar and rhetoric.[52] He backed away from the easy assumption of continuity between grammar and rhetoric found in Barthes, Genette and others, thinking that certain philosophical discrepancies and logical tensions lay concealed beneath such assumptions, that one ought at least to question the legitimacy of 'this reduction of figure to grammar'.[53] De Man wisely retreated from a theoretical exposition of this difficulty, opting instead to examine these tensions between literality and figurality in specific textual readings. Given that Adam Smith was both philosophical grammarian and rhetorician, de Man's framework promises a revealing exploration. It would begin simply, by putting back Smith's theory of language into its place of origin, the lectures on rhetoric.

The lectures on rhetoric took as their subject 'an examination of the several ways of communicating our thoughts by speech', and the 'principles of those literary contributions which contribute to persuasion or entertainment', in other words, discourse and style.[54] The theory of language and the consideration of figurality both occur early in the lecture sequence, as numbers three and six. The somewhat tentatively titled 'Of what is called the tropes and figures of speech' introduces the topic of figurality in an explicitly grammatical context, the ancient regulation of language by the rules 'we call Grammar'.[55] Figures were exceptions to these rules, so the Greeks and Romans, 'not candid enough to confess the grossness of their error and allow that these were exceptions', introduced a different kind of codification for these exceptions, under the name of tropes.[56] Figures first appear then as escapees from an original grammatical codification. Smith then proceeds to marginalise figures even further, as 'secondary mean sometimes proper to accomplish this end, to wit, when they more fitly expressed the sense of the author than the common stile'.[57] The numerous ancient and modern systems of rhetoric which devote themselves to figural language are a 'very silly set of Books and not at all instructive'.[58] Smith had some acute observations to make as he proceeded to discuss metaphor and metonymy, particularly on the difficulty of delimiting one figure from another, but generally speaking figural language did not receive a favourable press in Glasgow in late November, 1762. The overt reason appears to have been pedagogic. Elaborate and pervasive figuration went against the neat and elegant economy of style Smith was primarily recommending to his students. For our purposes, however, it is more important to notice firstly the evident tension between the categories of grammar and figure as the topic was broached, but equally the fact that, despite the progressive deprecation of the figural, grammar and figure were not necessarily opposing categories for Smith. Figure escaped grammar, not because of their irreconcilable essences, but because of the contingencies of human error. The category mistake of the ancients was to create a separate classification for the figural, instead of realising that the figural was simply the remainder left by their own original but imperfect grammatical regulation. The perfect grammar would presumably have included figure.

With this thought in mind we can return to what the essay on language teaches concerning grammar's beginning. It should now come as no surprise that the generative act of grammar is there openly characterised as figural. The necessity of common naming, which the savage is obliged to undertake if he is to have a workable language, receives the following description:

> We say, in the same manner, of a hero, that he is an Alexander; of an orator, that he is a Cicero; of a philosopher, that he is a Newton. This way of speaking, which the grammarians call an Antonomasia, and which is still extremely common, though now not at all necessary, demonstrates how much mankind are naturally disposed to give one object the name of any other, which nearly resembles it, and thus to denominate a multitude, by what was originally intended to express an individual.[59]

The kind of de Manian reading which would follow this admission needs only brief indication. Its premise would now be that grammar is originally figural. If this is so, then the acts creative of grammatical forms may be conceived as figuration. Generalisation, operating on the perception of similarity, produces a metaphoric mode, while abstraction, operating on separation, produces a metonymic. Since these are the poles between which figuration runs, grammatical forms are figures, not of themselves single metaphors or metonymies, but progressively complex combinations of these figural categories, because parts of speech require both sets of operation to come into being. The increasing difficulty of language, so emphasised by Smith under the name of metaphysics, would be re-read as the increasingly problematic nature of signification entailed by the indeterminate referential field which the figural perpetually creates. *I*, the linguistic sign of the self which is Smith's current limit of language's metaphysicality, would finally become an abysmally distant and irrecoverable figure.

This hypothetical reading is doubtless less adroit and perceptive than any which de Man himself could have produced, but now, sadly, cannot. Yet it may have the merit of reposing the problem his writing tackled. De Man saw the problem of the literal and the figural as one of tension, potential irreconcilability, and the interpretive undecidability induced by this condition. By contrast, there is discernible throughout Smith's understanding of language a quite stark and simple assertion of identities, not differences. Grammar is nature, rhetoric and metaphysics. The problem of reading Smith has been that his arguments could not simultaneously recognise all these identities. They are, as has been seen, by and large unconcealed in his various texts, but however much they are required on the various occasions of his argument they cannot be allowed to claim the same argumentative space, an irresolvable situation which Smith's writing in turn rhetorically proclaims. The problem of grammar and rhetoric is restated, through this particular consideration of Smith, as an identity which moves in and out of textual visibility, persisting in self-assertion towards eyes which must not see it. It is the assertion of identity itself which creates tension and irreconcilability, and is prior to them. The enquiry therefore becomes centred on an identity which can only be very partially acknowledged, and on the reasons for this situation, rather than on two categories opposed in their mode of signification.

Within the context of this essay, whose subject is Adam Smith, a contemporary theory of the relationship between the literal and the figural cannot be further pursued. By way of conclusion, it is more proper to reflect briefly upon the condition of that modern linguistic consciousness portrayed by Smith. Whether we name the perplexity he perceived in modern language as metaphysics or figuration is less important than the fact and nature of the perplexity. The speaker (or writer, or reader) of modern English exists in a state of linguistic estrangement from both nature and inward perception of that nature. The language he uses is unsexed and unlovely, constrained from the production of beauty by the necessity of rigidly ordered composition, and uttered by a linguistic self of dreadful complexity. As philosophy itself moves from Pascal to PASCAL, using machine intelligence in its attempts to understand the problems of, once again, natural and artificial grammar, Smith could surely have recognised the material apotheosis of the machine icon which delivered the linguistic plight and problems of the modern in the first place.

Notes

1 I have used the Glasgow edition of the works and correspondence of Adam Smith. The essays 'Of the external senses', 'The history of astronomy', 'The history of ancient physics', 'The history of ancient logics and metaphysics' and 'Of the imitative arts' are all in Adam Smith, *Essays on Philosophical Subjects,* eds. W. P. D. Brightman and J. C. Bryce, Oxford, 1980. 'Considerations concerning the first formation of languages' is available in Smith, *Lectures on Rhetoric and Belles Lettres,* ed. J. C. Bryce, Oxford, 1983.

2 Dugald Stewart, 'Account of the life and writings of Adam Smith', in Smith, *Essays,* p. 292.

3 *Ibid.,* p. 293.

4 *Ibid.*

5 Walter Bagehot, 'Adam Smith as a person', *Collected Works,* ed. N. St. John-Stevas, 13 vols., III, London, 1968, p. 91.

6 This is by some way the longest of Smith's early essays. It was certainly composed before 1758, and perhaps considerably earlier. See the editors' useful 'Introduction' to the essay in Smith, *Essays,* p. 7.

7 Smith, *Essays,* p. 105.

8 Smith, *Rhetoric,* Lecture 3, pp. 9–13.

9 Smith to George Baird, 7 February, 1763, quoted in the editor's 'Introduction' to Smith, *Rhetoric,* p. 24.

10 *Ibid.,* p. 25.

11 *Ibid.,* p. 24.

12 Smith, 'Considerations concerning the first formation of languages', in *Rhetoric,* p. 205. This essay will hereafter be referred to as 'Considerations'.

13 H. Aarsleff, 'The tradition of Condillac', in *From Locke to Saussure,* Minneapolis, 1982, and other relevant essays in this collection of Aarsleff's work.

14 Abbé de Condillac, *Essai sur l'Origine des Connoissances Humaines,* Paris, 1746; J. G. von Herder, *Uber den Ursprung der Sprache,* Berlin, 1771; J. Burnet, Lord Monboddo, *Of the Origin and Progress of Language,* Edinburgh, 1773–1792; J. J. Rousseau, *Essai sur l'Origine des Langues,* Paris, 1781.

15 J. J. Rousseau, *Discourse,* in *The Social Contract and Discourses,* ed. G. D. H. Cole, London, 1913 (repr. 1973), pp. 57–8.

16 Smith, 'Considerations', p. 203.

17 *Ibid.*

18 *Ibid.,* p. 204.

19 *Ibid.,* p. 205.

20 *Ibid.,* p. 204.

21 Rousseau, *Discourse,* p. 61.

22 *Ibid.*, p. 59.
23 *Ibid.*, pp. 57–63.
24 Smith, 'Considerations', p. 207.
25 *Ibid.*, p. 209.
26 *Ibid.*, p. 212–13.
27 *Ibid.*, p. 212.
28 *Ibid.*, p. 215.
29 *Ibid.*, p. 216.
30 *Ibid.*, 219.
31 *Ibid.*
32 *Ibid.*
33 *Ibid.*
34 *Ibid.*, pp. 207–8.
35 *Ibid.*, p. 211.
36 See, for example, John Webster, *Academiarum Examen,* London, 1653.
37 Cf. Bennington's essay in this volume, p. 110.
38 Smith, 'Considerations', p. 221.
39 *Ibid.*, pp. 221–2.
40 Smith: *Essays,* pp. 66–7; *Rhetoric,* p. 13.
41 Smith, 'Considerations', p. 223.
42 The only commentator I am aware of who entertains this approach is J. R. Lindgren, *The Social Philosophy of Adam Smith,* The Hague, 1973, chapter 1.
43 Smith, 'Considerations', p. 224.
44 *Ibid.*, p. 225.
45 *Ibid.*, p. 226.
46 *Ibid.*, p. 216.
47 *Ibid.*, pp. 207–8.
48 Smith, *Essays,* pp. 156–61.
49 *Ibid.*, p. 158.
50 *Ibid.*, p. 157.
51 *Ibid.*, p. 161.
52 Paul de Man, *Allegories of Reading,* New Haven, 1979, chapter 1.
53 *Ibid.*, p. 7.
54 Smith, *Rhetoric,* 'Introduction', p. 12.
55 *Ibid.*, p. 25.
56 *Ibid.*
57 *Ibid.*, p. 26.
58 *Ibid.*
59 'Considerations', p. 204.

What clarity this essay has owes much to the comments of my fellow editors, and to those of Dr. R. S. Porter and Dr. S. Shuttleworth, all of whose help I should like gratefully to acknowledge.